TRIVENT
PUBLISHING

TRIVENT

Ethics | in Science & Technology

Medical Futility in Paediatrics:
Interdisciplinary and International Perspectives

Edited by

Kartina A. Choong

With a Foreword by

Michael Redfern QC

Applied Ethics: From Bioethics to Environmental Ethics
Volume III

Medical Futility in Paediatrics: Interdisciplinary and International Perspectives
Ed. Kartina A. Choong
ISBN 978-615-6405-00-5
Applied Ethics: From Bioethics to Environmental Ethics, Volume III
Series ed. János I. Tóth
Responsible publisher: Teodora C. Artimon
ISSN 2630-8185
http://trivent-publishing.eu/ethics.html

First published in 2019 by Trivent Publishing

Trivent Publishing
1119 Budapest, Etele út 59–61
Hungary

For more information, visit our website: http://trivent-publishing.eu

TABLE OF CONTENTS

Foreword

This book addresses the disparate issues relating to the futility of further medical treatment on the one hand, and the treatment preferences identified by the families which doctors regard as having no therapeutic benefit. These issues present legal, medical, moral, ethical, and social problems. The editor and authors seek to analyse the interests of the parties including outcomes in comparative legal systems against the yardstick of early alternative dispute resolution as opposed to protracted litigation.

The cases of Charlie Gard and Alfie Evans in the UK and Jahi McMath in the USA raise critical issues. Each case was subjected to the full might, majesty, dominion, and power of the legal system in the UK, Europe, and the USA respectively. In the Charlie Gard and Alfie Evans cases, proceedings were commenced in the High Court, there was an appeal to the Court of Appeal, a further appeal to the Supreme Court, and finally a hearing in the European Court of Human Rights. In the case of Alfie Evans, this legal procedure was repeated more than once.

Each step along the way was poured over by the media and the press. There was an absence of privacy for the families. The stories attracted worldwide public interest. The extensive publicity could hardly be said to be in the best interests of the children involved. This factor alone can lead to a breakdown of trust between the parties.

The process throughout was adversarial, time consuming, and hugely expensive. The NHS' costs were in excess of £420,000. NHS funding requires reallocation of scarce resources to legal action which otherwise would have funded medical care. Legal Aid is a similarly scarce resource and not readily available. Disparity in funding can also exacerbate mistrust between the parties.

Families might have recourse to crowdfunding involving the promulgation of an appealing story in order to stimulate the funding of litigation. On this issue alone there is a balance to be struck between the best interests of the child and the significant harm argument. Family privacy is invaded by media and press attention. This invasion of privacy occurs at a time when privacy is vital in safeguarding the best interests of the children and their families.

There is much to commend the ability of hospitals and doctors on the one hand and families on the other to mediate prior to recourse to litigation. Mediation can draw on the work of medical ethics committees, legal, moral, social, and medical bodies in an effort to mediate and resolve the priorities of all the parties concerned in early course. This is infinitely preferable in most situations. Prolonged adversarial litigation is a huge burden and a source of ever-increasing levels of stress during the currency of the proceedings at a time when the converse should prevail. Recourse to protracted litigation should be a matter of last resort.

There must be the highest standard of medical diagnosis identifying medical futility and a fully reasoned response to what might be regarded as useless treatment. On the other hand, the realistic needs of families must be identified without their distress being fuelled by media frenzy and crowdfunding of legal costs. For this reason, although the recent changes introduced by the Mental Capacity (Amendments) Act 2019 may not be directly relevant for present purposes, the initiative to ascertain the wishes and feelings of patients who lack capacity signifies a positive attempt to improve family participation in decision-making about care and treatment whilst ensuring that those patients' welfare remains the core concern. Enhanced collaboration between families and healthcare professionals can minimise the risk of protracted legal disputes. This wind of change has considerable application in paediatric care.

The ability to resolve life's most difficult and sensitive medical treatment of children requires the utmost care and consideration. This book confirms that there is a lot to be said for transforming lives together when considering resource intensive treatment in a resource finite environment.

Michael Redfern QC

Preface

The cases of Charlie Gard and Alfie Evans were among the most high-profile doctors-parents disputes in England and Wales in recent times. They were contested throughout the entire hierarchy of the domestic court system, and beyond, to the European Court of Human Rights. Appeal after appeal failed to reverse the respective court of first instance's declarations that the continuation of ventilatory support and life-sustaining treatment were not in the infants' best interests. Still, what is particularly noteworthy about the two cases is not just the debates that took place in the courtrooms. Few medical law cases have attracted as much global attention and even interventions from well-meaning individuals and institutions unconnected to the two infants.

This book explores the issues and challenges posed by these and similar cases, both inside and outside the courtroom. It also examines how other jurisdictions would deal with comparable situations. The book consists of 17 chapters, written by scholars with expertise in law, medicine, medical ethics, theology, health policy and management, English literature, nursing, and history, from the UK, Australia, Canada, the Czech Republic, France, Germany, India, Spain, Turkey, and the USA.

The introductory chapter sets the context for the book by documenting the chronology of medical events relating to the two infants, as accompanied by the legal arguments which were fielded and debated during their protracted courtroom battles. The subsequent 16 chapters are organised into 2 Parts. Part I takes an interdisciplinary look at the wide range of issues raised in the two cases, as well as medical futility in paediatrics generally. Part II sheds light on how these situations are likely to be managed in other jurisdictions.

Part I begins with an important reminder from Jo Samanta that intractable disputes between doctors and parents relating to seriously ill infants are neither new, nor unusual. Her chapter examines the evolution of the best interests test when used in this area, and considers the possible role of mediation as a vehicle for alternative dispute resolution. William Seagrim's chapter defends the continued usage of the best interests test which, as a yardstick in paediatric cases, has been labelled the "gold standard." He points out that although some commentators have suggested that it insufficiently protects parents from state intervention and have called instead for an enabling "significant harm" threshold, such submissions are devoid of any legal basis. The following chapter by Sarah Sargent offers a contrasting viewpoint. In her discussion of *Charlie's Law*, she explores the significance and benefits of clearly establishing the "significant harm" threshold to ensure more protection for parental wishes and to reduce needless interference from healthcare professionals and the courts.

Looking next at the role of the media in making public the difficult nature of medical futility cases, Kim McGuire discusses interpretations of the media role, as both "positive" and "negative." Whilst for the families, media reporting widened access to avenues of support; for healthcare professionals, it brought abuse and declining trust. This chapter concludes by considering mediation as a means for expressing and resolving conflicting viewpoints outside the media gaze. Richard Law then identifies the extent to which resource-intense medical treatments place significant financial burdens on healthcare systems especially in the provision of life-sustaining care to critically unwell patients. His chapter analyses the clinical and legal implications of limited resources and the nature of the impact that medical crowdfunding exerts on resource allocation. The ensuing chapter by Alejandra Boto focuses on the reported conferral of Italian citizenship to Alfie Evans despite the apparent absence of an Italian lineage. Approaching the issue from a public law perspective, she discusses Italy's legal provisions relating to the conferral of citizenship to foreign nationals and provides a comparative analysis with Spain.

Brendan McCarthy's chapter highlights that the Church of England views medical ethics through the prism of four cascading principles: affirmation of life; care of the vulnerable; creation of a caring and cohesive society; and respect for individuals. This framework provides a principled, consistent guide for all areas of care including when determining the best interests of children in the context of potentially futile medical intervention. Mahmood Chandia and Abdulla al-Shami then look at the debate from an Islamic legal (*fiqh*) perspective by addressing two questions: what is the scope of parental rights and responsibilities in Islam; and how would Islamic *fiqh* guide parental thinking in such situations? Part I ends with Lisa Cherkassky's chapter which examines whether parents have the legal right to take their gravely ill hospitalised children home, to die, and whether this final wish should be afforded more weight in the future.

Part II begins with a chapter from Roslyn Jones which considers the legislative provisions and case law on comparable matters in Australia and analysing those cases in terms of issues of quality of life, professional obligations, parental versus state control, and the use or misuse of media and social media. Next, Helena Krejčíková highlights that in the absence of any explicit legislative recognition of medical futility in the Czech Republic, clinical practice relies predominantly on a vague statutory definition of proper care and two non-binding guidelines aimed, however, at adult patients. Her chapter introduces readers to the Czech legal framework and clinical practice, including a new document prepared by the Paediatric Palliative Care Working Group on providing life-sustaining care, which would apply in cases similar to those of Gard and Evans. French Law's response to medical futility relating to minor patients is then examined by Stephanie Rohlfing-Dijoux. Her chapter assesses how the country's end-of-life legislation can be conciliated with its child protection law and legislative provisions relating to minors in family law. Peter Elsner's chapter, which focuses on Germany, appraises how decisions regarding the continuation or termination of medical support measures in end-of-life situation involving child patients are balanced against the right to life enshrined in the German Constitution.

In the next chapter, Abhay Vaidya and Sourabhi Sahakari review the Indian healthcare and legal systems. By illustrating how different these are from the NHS and the standpoints taken by English Law, they seek to show that the Gard and Evans cases are unique products of their environment. Banu Buruk and Berna Arda explore in their chapter the different, and at times contrasting, set of challenges faced by doctors and parents in a communitarian society like Turkey, and describe how the best interests of child patients are usually determined therein. The final chapter in Part II by Vincent Maher compares Gard and Evans with the Jahi McMath case in the USA. It asserts that medical diagnoses pertaining to medical futility and treatment should take precedence over the unrealistic preferences of families fuelled by the dramatic pressures of social media.

The book hopes to enrich the ongoing debates surrounding medical futility in paediatrics, and to encourage parents and healthcare professionals faced with this dilemma in the future to consider a wider range of options.

I am grateful to the authors for their valuable contributions, and the peer reviewers for giving generously of their time to enhance the quality of the chapters. I would also like to thank Dr Teodora Artimon, the book's managing editor, for her enthusiasm, professionalism, and commitment to this project.

Kartina A. Choong
University of Central Lancashire, UK
July 2019

Notes on Contributors

Abdulla Al-Shami is a Professor of Arabic and Islamic Studies who is now a Senior Research Fellow at the Tessellate Institute, Canada. He obtained his LIB in Islamic Law (Shari'ah) and ILM degree in Comparative Islamic Jurisprudence from Saudi Arabia, and a PhD in Comparative Islamic Jurisprudence from the University of Manchester. He has published a total of 6 books and over 60 research articles on Islamic law and Islamic studies. His research focuses on Islamic banking and finance, Islamic governance, political Islam, Islamic law (family, criminal, contract), Qur'anic and Hadith Studies, Islamic law and human rights, Islamic professional and medical ethics.

Berna Arda is a Professor in the Department of Medical Ethics and History of Medicine at the Faculty of Medicine, Ankara University. She is a graduate of the same Faculty and holds medical specialty and PhD degrees. Her main research and publication fields are science ethics, human rights, women and bioethics, medical law, ethics education and disease concept in the history of medicine. She was a visiting scientist at the Boston Children's Hospital and Harvard Medical School in 2015, and a guest professor at University College London's History of Medicine Centre in 2008. She was the founder chairperson of the Turkish Bioethics Society, and a member of the High Disciplinary Committee of the Turkish Medical Association. She is also the governor and treasurer of the World Association for Medical Law, and was the founder and first president of the International Association for Education in Ethics.

Alejandra Boto is a Senior Lecturer at the University of Oviedo and she holds a PhD in Administrative Law. Her research interests cover administrative organisation, public services provision and users' rights. She is the author of 3 monographs and several parts of books published in Canada, Germany and Spain on a wide range of public law topics. In 2016, during the XXV Jurists of Health Conference, she received a special prize awarded by the Spanish Society of Public Health and Health Administration for a communication regarding standardised packaging of tobacco products. Recent works also include several articles in journals published in the United States, France, Slovenia, Italy and Colombia, focusing on comparative law.

Banu Buruk has been a Project Manager at the Turkish Research Funding Agency (TUBITAK) for the last 12 years. She teaches at the Department of Medical Ethics and History of Medicine in different universities in Ankara and gives seminars on research management and research ethics in different academic environments. She is a graduate of the Department of Molecular Biology and Genetics of Istanbul Technical University and has a PhD degree in Medical History and Ethics from the Faculty of Medicine at Ankara University. Her main research and publication fields are research ethics, nanoethics, ethics of personal data usage in research and ethics education. She visited Boston Children's Hospital and Harvard Medical School in 2014 as an NIH-Fogarty Fellow, and the Western Institutional Review Board (WIRB) in 2018. She is a member of the Clinical Research Ethics Committee of TOBB ETU University and a member of the National Open Science Committee since 2017.

Mahmood Chandia is a Senior Lecturer in Religion, Culture and Society & Volunteering and Community Leadership at the University of Central Lancashire. He studied Islamic Studies at Darul Uloom, Bury, and at the Al-Azhar University, Cairo. He holds a Masters degree in Middle Eastern Studies from the University of Manchester. His PhD in Middle Eastern Studies was also obtained from the University of Manchester. His areas of research interests are wide-ranging. He has published on the following topics: medical futility, religiously- and culturally-sensitive end-of life-care, the Halal food chain, religious and legal understandings of Halal slaughter, suicide and Islam, and ethnicity and religion. He has also produced a complete translation of the Holy Quran in both digitised and print formats.

Lisa Cherkassky is a Senior Lecturer in Medical Law at the University of Derby and a Visiting Professor of Comparative Medical Law and Ethics at Western University, Ontario. She studied Law at the University of Sunderland (LLB and LLM) before taking post as a Lecturer in Medical Law at the University of Bradford in 2007. Her main research areas are saviour siblings, liver transplantation, the welfare of the child, preimplantation genetic diagnosis, posthumous gamete retrieval, and the autonomy of the parents in medical situations. She has also published on reckless HIV transmission, causation, and genocide.

Peter Elsner is a Professor and Chairman of the Department of Dermatology, University Hospital Jena. He was trained at the Medical School of Julius-Maximilians-University of Wuerzburg and is a board-certified dermatologist and allergist. He has previously worked at the Department of Dermatology, UCSF, San Francisco and at the Dermatology Department of the University Hospital Zurich. He has published over 600 scientific papers and is the author of more than 20 books. He holds an LLB from the University of Central Lancashire. In addition to his research in contact dermatitis and dermatoepidemiology, he has worked and published on medical liability and medico-ethical issues.

Roslyn E. Jones graduated in Medicine in 1980 from the University of New South Wales. Following a period of clinical work, Roslyn qualified in a Masters of Health Planning from the University of New South Wales in 1985 and was then admitted as a Fellow of the Royal Australian College of Medical Administrators in 1990. She has worked in administrative roles in health policy with the New South Wales Department of Health and Director of Medical Services positions in public hospitals in Sydney. Roslyn's current position in medico-legal management in a Sydney teaching hospital involves the difficult intersection between medicine and law, and affords her an appreciation of the strengths and limitations of each.

Helena Krejčíková is a Member of the Institute for Criminal Law, Police Academy of the Czech Republic in Prague, and a Scientific Researcher at the Centre for Medical Law, Law Faculty, Charles University in Prague. She also works as a Scientific Researcher and a Legal Expert of the Honorary Board of the Czech Medical Chamber. She is a member of the Czech Society for Palliative Medicine and was one of the drafters of the Czech Paediatric guidelines. Her main research interests lie in end-of-life decisions and the criminal liability of doctors and healthcare providers.

Richard W. M. Law is an Associate Specialist in Acute Medicine at the University Hospitals Morecambe Bay NHS Foundation Trust and works at the Royal Lancaster Infirmary. He studied Medicine at the University of Leeds before qualifying as a General Medical Council-registered medical practitioner. He is currently a practising physician in acute internal medicine. He holds a Masters (LLM) degree in Law from the University of Central Lancashire. His main research interests in relation to clinical practice are medical futility, decision-making in the context of withdrawal of life-sustaining treatment, and resource allocation decisions in relation to critical care.

Vincent F. Maher is a Professor of Management and Health Care Management at the LaPenta School of Business of Iona College, New York. He received his Law degree from the Law School of the City University of New York, his Masters in Nurse Anaesthesia and Health Administration from Columbia University and was awarded a certificate in comparative ethics by Girton College, University of Cambridge. He has taught an array of management, ethics and law classes throughout his 30+ years in academia. He also teaches nursing to undergraduate and graduate nursing students. He is a licensed attorney who practised medical negligence law on behalf of plaintiffs, an ethicist and poet, and a retired certified registered nurse anaesthetist with specialisation in shock, trauma and critical care. He has written and lectured extensively in the fields of law, medicine, healthcare ethics and decision-making and is a Fellow of the New York Academy of Medicine.

Brendan McCarthy is the National Adviser on Medical Ethics and Health and Social Care Policy for the Church of England. Previously, Brendan had been a Church of Ireland rector and hospital chaplain and then chief executive of a cross-community Christian charity in Northern Ireland. He has also been a non-executive director of the Western Health and Social Care Trust (NI) and a Board member of the Northern Ireland Practice and Education Council for Nursing and Midwifery.

Kim McGuire is a Senior Lecturer in Law at the University of Central Lancashire. She studied History and English Literature at Lancaster University, and Law at the University of Central Lancashire. She holds Masters degrees in Law (University of Central Lancashire) and in Social History (Lancaster University), and a PhD in History (Lancaster University). Her main areas of research interests, on which she has published extensively, include inter-personal conflicts in various social and economic contexts. The foci are conflict construction (linguistically, legally, emotionally, factually), participants' perceptions of self and others, and ultimately how contentions can be perpetuated, mediated, regulated and resolved.

Stephanie Rohlfing-Dijoux is a Professor of Law at the University Paris Nanterre since 2008 and a Lawyer at the Paris (France) and Munich (Germany) Bars since 1994. She holds a PhD in German Law from the University Frankfurt/Main and in Civil Law from the University Paris I Sorbonne. She is a Director of the German-French integrated law curriculum programmes of the German-French University for the Bachelor, Masters and PhD Levels, and is the co-director of the Master II Bilingue des droits de l'Europe. She is a member of the multidisciplinary and multi-language research laboratory (CRPM) at the University Paris Nanterre. Her main areas of research interest are culture and law, comparative law and medical law.

Sourabhi A. Sahakari was previously an Assistant Professor of Law at the Govind Ramnath Kare College of Law, where she taught modules on women and law, family law, interpretation of statutes, human rights and banking law. Prior to that, she worked as an attorney at Exactus Corporation Private Ltd for a period of one year. She holds a Masters degree in Law (Criminology) from Goa University and graduated at the top of her class. She is currently pursuing a PhD in Law on the topic of victim compensation and rights at the Govind Ramnath Kare College of Law, in addition to working as a classical dance (Bharatnatyam) teacher for the Government of Goa's Directorate of Art and Culture.

Jo Samanta is Emeritus Professor of Medical Law at De Montfort University. She practised as a company commercial solicitor before becoming an academic. She also qualified and practised as a nurse and midwife in South Africa before returning to the UK. She has published extensively in leading legal and medical journals, and has a particular interest in empirical interdisciplinary studies. Her current areas of research interest include end-of-life decisions, clinical guidelines and medical innovation.

Sarah Sargent is a Reader in Law at the School of Law, University of Buckingham, where she lectures in family and medical law. She obtained her PhD from De Montfort University, which focused on the interpretation of the "best interests" of the child standard in intercountry adoption. She also holds an LLM in Social Welfare Law from the University of Leicester. She obtained her Juris Doctor degree from the University of Denver. Her research and publication interests include children's issues, including the interpretation of children's rights.

William Seagrim is a Lecturer at Cardiff University and a Barrister at 9 Park Place Chambers, Wales. He studied Law at Swansea University and obtained a first class honours degree. William then qualified as a Barrister (Honourable Society of Gray's Inn, London) and completed a common law pupillage at 9 Park Place Chambers. He has over 11 years' experience at the Bar of England and Wales and specialises in family law. His practice encompasses all areas of family law, including disputes regarding the medical treatment of children. His teaching and research interests include family law, professional ethics and advocacy.

Abhay Vaidya trained as an Anaesthetist in India, and has worked as a Consultant Anaesthetist and Intensivist in India, the Gulf countries and the UK. He is a Fellow of the Royal College of Anaesthetists, the Higher Education Academy and a member of the Academy of Medical Educators. As an intensivist, he was often involved in the decision-making processes that considered the best interests of patients especially when deciding the withdrawal and/or withholding of treatment in the adult intensive care unit. He is currently a Small Group Facilitator for medical law and medical ethics at the School of Medicine, University of Central Lancashire.

INTRODUCTION

Charlie Gard and Alfie Evans: Their Medico-Legal Journeys

Kartina A. Choong
University of Central Lancashire, UK

I. Introduction

Charlie Gard and Alfie Evans were two desperately ill children born to young first-time parents. They were treated at two renowned publicly-funded paediatric hospitals in England, namely the Great Ormond Street Hospital for Children NHS Foundation Trust (GOSH) in London and the Aldey Hey Children's NHS Foundation Trust (Alder Hey Hospital) in Liverpool. The cases first captured public attention when both sets of parents vehemently opposed the court declarations sought by the hospitals that it would be in the infants' best interests to have their life-sustaining treatment (LST) withdrawn after the treating medical teams decided that the continuation of LST was futile and may even lead to the prolongation of pain and suffering. In each case, the declaration was granted by the High Court. In both, the parents appealed to the highest court in the land as well as to the European Court of Human Rights (ECtHR) in Strasbourg. However, none of the appeals managed to overturn the initial court declarations, and both infants died shortly after their LST was discontinued. This context-setting chapter chronicles their medico-legal journeys[1] and concludes by reflecting on their legacies for similar cases in the near future.

II. Charlie Gard

Born on 4th August 2016, Charlie Gard[2] appeared healthy at birth. He gave cause for concern in the first two months when he seemed less

[1] The background information and medico-legal details referred to in this chapter are derived from the case reports cited herein. Where necessary, reference will be made to the relevant pages or paragraphs of those judgments.
[2] Hereinafter referred to as "Charlie."

able to lift his head and support himself compared to other babies his age. He was also not gaining weight. On 11th October 2016, he was placed under the care of GOSH after he became increasingly lethargic and his breathing became shallow. He was diagnosed as suffering from infantile onset encephalomyopathic mitochondrial DNA depletion syndrome (MDDS). This very rare and debilitating condition is caused by biallelic mutations in a gene called RRM2B, whose effect was to leave Charlie with severe depletion of the amount of mitochondrial DNA in his tissues. This had a profound impact upon his brain, muscles, and ability to breathe, and some effect on his heart, liver, and kidneys. As a consequence, he had progressive muscle weakness including of his breathing muscles. He was also not able to move his arms, fingers, legs, and toes; and not always able to open his eyes. As part of this underlying condition, he was also profoundly deaf. He further developed a severe epilepsy disorder and suffered persistent seizures. Although he was not brain dead, he did not display the usual signs of normal brain activities like responsiveness, interaction, or crying. Owing to the severity of his condition, regarded as "exceptionally rare,"[3] he had been, since his arrival at the hospital, cared for in the intensive care unit (ICU) where he breathed with the assistance of mechanical ventilation.

His parents, Connie Yates and Chris Gard, in the course of their researches came across a form of treatment known as nucleoside therapy which they had hoped would help decelerate or even halt Charlie's relentless deterioration. This therapy had evidently been beneficial when used on MDDS patients with TK2 mutation, a different and less severe mitochondrial condition that primarily causes muscle weakness but does not generally affect the brain. It operates by introducing an alternative source of energy which the cells can utilise to replace the energy depletion caused by MDDS. The energy is supplied through a chemical compound which is added to the patient's food. The parents made contact with a US-based Professor of Neurology with relevant experience of administering the therapy. He cautioned that there was no direct evidence that the therapy had any beneficial effect on the brains of TK2 patients, and that nucleoside therapy had up to then

[3] *Great Ormond Street Hospital v Yates, Gard & Gard* [2017] EWHC 972 (Fam), para. 15.

never been experimented on animals or human beings with RRM2B deficiency. It was therefore unknown whether the energy-giving supplement would replenish brain cells; whereas in order to be helpful to patients with RRM2B, it would have to cross the blood/brain barrier, something which he opined was a theoretical possibility.

When this option was posed by the parents to the doctors at GOSH, they briefly contemplated it in early January 2017 and began to prepare the documentations for Ethics Committee approval as the treatment was then still experimental. However, following an episode of brain seizures on 9th or 10th January 2017 which continued intermittently for 3 weeks, Charlie's doctors came to the conclusion that nucleoside therapy would be futile. The plan was therefore abandoned. However, as the parents were still hopeful that the nucleoside therapy would offer some amelioration of Charlie's condition and that the administration of the supplement would cause little or no harm, they turned to crowdfunding to finance Charlie's journey to, and treatment in, the USA.[4]

Charlie's medical team, in the meantime, was of the opinion that the infant was still capable of experiencing pain. Hence not only would the administration of nucleoside therapy be potentially painful and could lead to the prolongation of his suffering, but that being ventilated, suctioned, and living as he did, were all possibly pain-inducing. They came to the decision that his condition had reached the stage where artificial ventilation should be withdrawn and only palliative care be given so as to allow him to die in peace and with dignity.

- High Court[5]

Thus on 24th February 2017, GOSH applied to the High Court for orders to be made (a) that Charlie lacked capacity to make decisions relating to his medical treatment; and (b) that it was lawful and in his best interests for his artificial ventilation to be withdrawn; and (c) for him to be provided only with palliative care. The application, made pursuant to the inherent jurisdiction of the court, was opposed by Charlie's parents. They wanted GOSH to keep Charlie ventilated and

[4] It was reported that they managed to gather approximately £1.3 million for this purpose – see C. Dyer, "Law, Ethics, and Emotion: The Charlie Gard Case," *British Medical Journal* 358 (2017): j3152, doi:10.1136/bmj.j3152.
[5] *Great Ormond Street Hospital* (n 3).

alive until they could take him to the USA. As the question of possible nucleoside therapy was raised by the parents, GOSH added to their application for an order that (d) it would not be in Charlie's interest to undergo nucleoside therapy. Mr Justice Francis, who heard the case, joined Charlie to the proceedings and appointed a guardian to represent his interests.

Given his age, Charlie's lack of capacity was plainly not in issue. Outlining the relevant legal test for the other three matters, Mr Justice Francis acknowledged that whilst parents with parental responsibility have the power to make decisions regarding their child's medical treatment, overriding control is vested in the court. The latter exercises its independent and objective[6] judgment in the child's best interests. For this, the court is guided by the principle that the child's welfare shall be its paramount consideration, as enshrined in section 1 of the Children Act 1989.

In determining Charlie's best interests, Mr Justice Francis heard both the testimony of the treating team at GOSH and other eminent doctors from the UK including one who was instructed by the parents to give evidence. Evidence was also received over the telephone from the aforementioned American Professor of Neurology, whose identity we now know as Professor Michio Hirano[7] of Columbia University. All agreed that the nucleoside therapy would offer no effective benefit to Charlie and further treatment would be futile. The word "futile," Mr Justice Francis clarified, means "pointless or of no effective benefit."[8] Even Professor Hirano, having been presented with Charlie's medical records, said that "I agree that it is very unlikely that he will improve with that therapy. It is unlikely."[9] At most it would make a "modest difference in life expectancy,"[10] but could certainly not undo structural

[6] Gillon nevertheless questioned whether it is ever possible to make an "objective" judgment in acute moral dilemmas such as this – see R. Gillon, "Why Charlie Gard's Parents Should Have Been the Decision-Makers About Their Son's Best Interests," *Journal of Medical Ethics* 44 (2018): 462-465, at 464.

[7] S. Mayor, "Charlie Gard Dies in Hospice After High Court Rules Against Prolonging Life Support," *British Medical Journal* 358 (2017): j3673, doi:10.1136/bmj.j3673.

[8] *Great Ormond Street Hospital* (n 3), para. 93.

[9] Ibidem, para. 98.

[10] Ibidem, para. 106.

brain damage. He placed the benefit of the treatment to Charlie's brain "as low, but not zero"[11] and claimed that had Charlie been in the USA, he would have treated him.[12] In words that seemed to convey compassion rather than hope, he said "I would just like to offer what we can. It is unlikely to work, but the alternative is that he will pass away."[13]

However, given that Charlie's brain damage was already severe and irreversible, and nucleoside therapy was potentially painful and would not achieve anything positive,[14] Mr Justice Francis held, in a judgment dated 11th April 2017, that it was in Charlie's best interests "to let him slip away peacefully and not put him through more pain and suffering."[15] He therefore acceded to GOSH's application and made formal declarations that it was not in Charlie's best interests to continue to be artificially ventilated; and that it was not in his best interests to undergo nucleoside therapy in the USA. It was instead in Charlie's best interests to have the artificial ventilation withdrawn and for him to be provided only with palliative care; provided that the measures taken are compatible with maintaining Charlie's dignity. The judge ended his ruling with the recommendation that in future, some form of Issues Resolution Hearing or mediation could be attempted for similar cases in the hope that the confidential conversations between the parties could help them find common grounds, or a resolution, or at the very least, a better understanding between them.[16]

- Court of Appeal[17]

On 2nd May 2017, the parents sought permission to appeal against the High Court's declarations, and to obtain declarations indicating that they had the right to transfer Charlie to a clinic where he could be provided with nucleoside therapy. GOSH agreed to maintain Charlie's treatment pending a full hearing. Of the five potential grounds of appeal lodged, only three were accepted. These relate to contentions that: a different approach should be employed as a matter of law to cases

[11] Ibidem, para. 104.
[12] Ibidem, para. 99.
[13] Ibidem, para. 127.
[14] Ibidem, para. 17.
[15] Ibidem, para. 128.
[16] Ibidem, para. 130.
[17] *Yates & Gard v Great Ormond Street Hospital & Gard* [2017] EWCA Civ 410.

where a choice has to be made between viable treatment options; the High Court judge had no jurisdiction to grant an order on the application of one clinical team preventing another clinical team from carrying out a course of treatment that was offered in the reasonable exercise of professional judgment; and no or insufficient regard was given to Charlie's and his parents' rights under the European Convention on Human Rights (ECHR) (but permission to appeal was confined only to human rights arguments that supplement the previous two grounds).[18]

Concerning the first ground of appeal, the parents argued that the judge erred in making an order which prevented Charlie from being removed and transported to a reputable hospital abroad in circumstances where there was no risk that the treatment would cause "significant harm" to him. For this, they argued that it is important that a distinction is drawn between two categories of cases. "Category 1" refers to cases where parental opposition to the course of action proposed by the medical team is not supported by a viable alternative therapeutic option. "Category 2," by contrast, refers to cases where a viable alternative treatment option is presented by the parents. In such situations, it was contended that the parents' choice should only be overridden by the courts if it is established that the alternative route is likely to cause the child to suffer "significant harm." Whilst "category 1" cases should continue to be decided by reference to the "best interests" test, the legal decision-making process for "category 2" cases must be structurally different in order to safeguard the parents' autonomy and ability to pursue their choice of treatment insofar as it does not pose "significant harm" to the child. They cited *Re King*[19] in support, where the court, in ruling that the parents' wish to take their child abroad for a viable alternative treatment (namely proton beam therapy) was "entirely reasonable,"[20] stated that the:

> responsibility for making decisions about a child rest with the parents. In most cases, the parents are the best people to make decisions about a child and the state – whether it be the court, or any other public authority – has no business interfering with the exercise of parental responsibility

[18] Ibidem, para. 37.
[19] [2014] EWHC 2964 (Fam).
[20] Ibidem, para. 34.

> unless the child is suffering or is likely to suffer significant harm as a result of the care given to the child not being what it would be reasonable to expect a parent to give.[21]

In light of this, Connie Yates' and Chris Gard's legal team argued that the High Court had erroneously applied the best interests test (which is reserved for "category 1" cases) to their situation. The imposition of the withdrawal of LST upon the child of parents who hold a legitimate opposing viewpoint favouring a viable treatment option which does not cause "significant harm" constituted, they further advanced, an intervention by the state on their right to private and family life as enshrined by Article 8 of the ECHR. Articles 2 and 5 were also mentioned in their submission, but these were not developed.

Next, their legal team acknowledged that GOSH was indeed entitled to apply for declarations concerning the lawfulness of its own plan to withdraw treatment, provide only palliative care and withhold nucleoside therapy on its site. However, the same cannot be said for its application to actively prevent external parties from arranging and delivering nucleoside therapy offsite. This, they claimed, was plainly beyond GOSH's powers as a public authority and/or the court's jurisdiction where proof of "significant harm" was not present. It was thereby wrong for the hospital to impose its view upon the parents by making the application; and the court, correspondingly, had no jurisdiction to uphold and support the hospital's stance on this matter. They were particularly concerned that the judge's declaration had *de facto* injunctive effect in that it prevented Charlie from being removed from GOSH to undergo treatment in the USA.

To these, the Court of Appeal responded as follows. As the first submission was premised upon the existence of a viable alternative treatment option, the fact that Mr Justice Francis had taken a very dim view of nucleoside therapy's potential to help Charlie, in and of itself signified that the question of whether, as a matter of law, there is a group of cases labelled "category 2" did not arise for Charlie. On the second argument that GOSH was acting outside its legal powers in bringing the application regarding nucleoside therapy, the Court of Appeal reminded the parents that it was they, and not the hospital, who raised the issue. And since Professor Hirano had offered to treat Char-

[21] Ibidem, para. 31.

lie in the USA by the time of the High Court hearing, GOSH raised it for determination within its overall application. Hence, rather than GOSH seeking to force its views on the parents and thereafter seeking the court's backing, it fell upon the judge to decide on the matter and this was reached on the basis of Charlie's best interests. It was thus incorrect on the part of the parents' legal team to frame this issue as being driven by GOSH. With this, the Court of Appeal concluded by a judgment dated 23rd May 2017 that both grounds of appeal, as supported by human rights arguments made under the third ground of appeal, must be dismissed; and that the High Court's ruling stood.

- Supreme Court[22]

The parents then lodged an appeal to the Supreme Court but their application for permission to appeal, presided over by Lady Hale, Lord Kerr and Lord Wilson, was declined on 8th June 2017. The Supreme Court affirmed that the High Court judge had applied the correct test, namely the "best interests test," in deciding Charlie's case and it also endorsed the factual findings of the High Court judge. It maintained that parents are not entitled to insist on treatment options which are not in their child's best interests and that the significant harm requirement does not apply to hospitals seeking guidance on their patients' best interests.

Not willing to give up, the parents took their case to the European Court of Human Rights (ECtHR) on 19th June 2017. The application was made by themselves and on behalf of Charlie. To facilitate this, the UK government requested that the Supreme Court consider making an order for a short stay of the declarations dated 11th April 2017. This was with a view to enabling the ECtHR to determine the parents' request under Rule 39 of the Rules of the Court, and to indicate to the United Kingdom any interim measures that should be adopted or for the proper conduct of the proceedings. Recognising how a further stay would require GOSH to provide artificial ventilation when this had already been declared as not in Charlie's best interests, the Supreme Court noted that

> [w]e three members of this court find ourselves in a situation which, so far as we can recall, we have never previous-

22 *In the Matter of Charlie Gard* (Supreme Court, 8th June 2017).

> ly experienced. By granting a stay, even of short duration, we would in some sense be complicit in directing a course of action which is contrary to Charlie's best interests.[23]

The court nevertheless directed that Mr Justice Francis' declarations be further stayed for three weeks until midnight on 10th/11th July 2017. This they did "with considerable hesitation,"[24] whilst urging the ECtHR to address the proposed application within that period, since they "would feel the gravest difficulty if asked to act yet further against Charlie's best interests by directing an even longer extension of the stay."[25]

- European Court of Human Rights[26]

Before the ECtHR, the parents complained both on their behalf and on Charlie's behalf under Articles 2 and 5 of the ECHR. As regards Article 2, it was their argument that GOSH had blocked LST and that this constituted a violation of the positive obligation under the Article. This had resulted in Charlie being deprived of his liberty within the meaning of Article 5. They also put forward complaints on their own behalf under Articles 6 and 8. Their grievance as regards Article 6 related to how the Court of Appeal came to the conclusion that their intended parental decisions would cause Charlie "significant harm" without the benefit of witness evidence. As for the Article 8 violation, it was argued that the High Court's declarations in April 2017 and other domestic courts' decisions formed a disproportionate interference with their parental rights. By taking their decisions in Charlie's best interests, rather than through an examination of whether there was a likelihood that Charlie was suffering or was likely to suffer significant harm, this constituted a disproportionate and unjustified interference with their parental rights under Article 8.

The parents did not provide reasons as to why the court should consider that they have standing to complain on Charlie's behalf under Article 2. To make this finding on standing, the court had to therefore satisfy itself, using established criteria, that (a) there was a risk that

[23] *In the Matter of Charlie Gard* (19th June 2017), para. 17.
[24] Ibidem, para. 20.
[25] Ibidem.
[26] *Gard and Others v United Kingdom* (application number 39793/17).

Charlie, as the direct victim, would be deprived of effective protection of his rights; and (b) there was an absence of a conflict of interest between Charlie and his parents. As regards (a), the fact that Charlie had been and continued to be represented by a court-appointed guardian whose task it was to ensure that Charlie's voice could be heard, led the court to the conclusion that the risk was minimised as far as possible. The court next considered (b) whether there was a conflict of interest between Charlie and the parents that would severely compromise the latter's eligibility to make an application on Charlie's behalf. Taking into consideration the "unambiguous and repeated findings"[27] of the domestic courts that the route they sought was not in Charlie's best interests, the court concluded that this was a clear conflict of interest between the parents and Charlie. It was therefore held that the parents did not have standing to make an application under Article 2 in Charlie's name and on his behalf.

The parents, on their own behalf, averred that GOSH had, through the domestic legal proceedings, blocked access to LST for Charlie. On this point concerning access to experimental treatment, Article 2 could not, according to the ECtHR, be interpreted as requiring access to unauthorised medicinal products for the terminally ill to be regulated in any particular way. This aspect of their complaint was therefore considered as manifestly ill-founded. Similarly for Article 5, whereupon the court pointed out that the availability of a domestic legal framework and the possibility to apply to the domestic courts, as creating the necessary safeguards regarding detention or deprivation of liberty. Hence, a claim under Article 5 was also considered as manifestly ill-founded.

As their Article 6 complaint related to how domestic courts arrived at their rulings, and this concerned an alleged interference in their private and family life, it was argued as a supplemental aspect of the parents' Article 8 complaint. It was therefore examined under Article 8 alone. In the light of previous case law on similar matters, the court acknowledged that there was an interference in the parents' Article 8 rights. Such an interference is a violation of the Article unless, pursuant to Article 8(2), it is in accordance with the law and is necessary in a democratic society. Considering the latter, the court opined that the hospital should indeed approach the court when conflicts arise, and

---

[27] Ibidem, para. 67.

that the UK holds and engaged the appropriate legal framework. Rather than being arbitrary and disproportionate, domestic courts have been meticulous and thorough, heard and considered the opinions of numerous expert witnesses, had their decisions reviewed at 3 levels of jurisdiction and applied a test (namely the best interests test) which had been endorsed at domestic and international levels as the correct test in such cases. Their application on this ground was also thereby held to be manifestly ill-founded.

The ECtHR sat on 27th June 2017 and declared the application inadmissible the next day on 28th June 2017.

- High Court[28]

On 7th July 2017, the case returned to the High Court, upon the application of GOSH. The hospital asked the court to affirm the declarations it made on 11th April 2017 but to express them in the form of orders which would then be enforceable. This, the hospital felt, would provide the clarity which declarations do not hold.

Mr Justice Francis noted that the case had by then escalated to an international scale and had even involved prominent world leaders.[29] He observed that "[a] lot of things have been said, particularly in recent days, by those who know almost nothing about this case but who feel entitled to express opinions. Many opinions have been expressed based on feelings rather than facts."[30] Although social media has, he opined, "very many benefits, one of its pitfalls is that when cases such as this go viral, the watching world feels entitled to express opinions, whether or not they are evidence-based."[31]

This application was unopposed. According to Mr Justice Francis,

> [t]he parents have had to face the reality, almost impossible to contemplate; that Charlie is beyond any help even from experimental treatment and that it is in his best interests for him to be allowed to die. Given the consensus that now exists between parents, the treating doctors and even Dr Hirano, it is my very sad duty to confirm the declarations

[28] *Great Ormond Street Hospital v Yates, Gard & Gard* [2017] EWHC 1909 (Fam).
[29] Ibidem, para. 10.
[30] Ibidem, para. 1.
[31] Ibidem, para. 11.

> that I made in April this year, and I now formally do so. I do not make a mandatory order.[32]

He also renewed his recommendation that mediation be used in all such cases in the future even if the only thing to come out of it is a better understanding among the parties.[33]

Charlie died on 28th July 2017,[34] a few days before his first birthday.

III. Alfie Evans

Alfie James Evans[35] was born on 9th May 2016, i.e. three months before Charlie Gard. Like Charlie, he seemed healthy and well at birth. At two months old, however, he was referred to hospital for a "divergent squint." Then, at four months old, during his first child development check with the General Practitioner, Alfie's mother intimated that she had concerns about his general development. He was smiling less frequently; not interacting; and sleeping far longer than had seemed normal to the point where she would commonly had to wake him up. Furthermore, he showed little inclination to play with his toys. At six months, Alfie was clearly displaying significant developmental delay. He was assessed as functioning in a range appropriate to a six-week to a two-month old child. An MRI brain/spine scan performed on 30th November 2016 suggested either a degenerative or metabolic disorder. In December 2016, he was taken by his parents to the Accident and Emergency Department of the Aldey Hey Hospital after an episode of rhythmic jerking of all four limbs and his jaw, alongside a number of other problems. He was transferred to the Paediatric Intensive Care Unit (PICU), where he was observed as being profoundly encephalopathic/comatose and unresponsive to painful/uncomfortable stimulation. He was ventilator-dependent and it was clear that he had suffered abnormal loss of muscle control.

Further scans conducted in February 2017 and August 2017 showed that 70% of his brain had been destroyed. After this third scan, the

---

32 Ibidem, para. 14.

33 Ibidem, para. 20.

34 BBC News, "Charlie Gard Parents Announce Death of 'Beautiful Boy'" (28th July 2017), https://www.bbc.co.uk/news/uk-england-london-40752120 (accessed April 7, 2019).

35 Hereinafter referred to as "Alfie."

hospital entered into discussion with Alfie's parents and suggested that his treatment be limited including the insertion of "Do not resuscitate" in Alfie's notes. Although the parents were initially receptive to these ideas, they subsequently changed their minds. Opinions from external experts were sought. These included doctors from Munich and the Bambino Gesù Hospital in Rome who travelled to the Alder Hey Hospital to examine Alfie. Whilst they agreed that Alfie had an untreatable and incurable neurodegenerative disease, they intimated that their hospitals were willing to treat Alfie at their premises.

Alder Hey Hospital issued an application to the High Court on 5th December 2017 for a declaration from the court that it was not in Alfie's best interests to receive ongoing ventilatory support and that it was unlawful for this to continue. On 8th January 2018, a mediation meeting took place but it was reported that no agreement was achieved.

- High Court[36]

The case came before Mr Justice Hayden in the High Court in February 2018. The hospital's application was opposed by Alfie's parents, Tom Evans and Kate James, who insisted that ventilation should be continued for the purposes of transporting him by air ambulance to the Bambino Gesù hospital in Rome, and subsequently perhaps to a hospital in Munich. He was supposed to undergo a tracheostomy and a gastronomy at one of these hospitals so as to enable ventilation to be given in a home environment (in one of these countries). If no improvement was to be registered after six months, they would accept the withdrawal of LST and Alfie would be allowed home to die.

As in the case of Charlie Gard, Alfie was represented by a guardian during the trial and the judge heard expert testimony from Alfie's treating clinicians, and external experts including those commissioned by his parents. The opinions offered were unanimous in that Alfie's brain condition was irreversible. However, unlike Charlie whose diagnosis had never been in any doubt, none of the doctors were able to offer a firm diagnosis for Alfie's condition.[37] There was nevertheless a consen-

---

[36] *Aldey Hey Children's NHS Foundation Trust v Evans, James & Evans* [2018] EWHC 308 (Fam).

[37] It seemed that this was not known until after Alfie's death, where it was reported that he had been suffering from the extremely rare brain condition GABA-transaminase deficiency – see BBC News, "Alfie Evans Parents

sus among all the doctors consulted that Alfie was suffering from a severe and progressive neurodegenerative disorder that was both catastrophic and untreatable.

Mr Justice Hayden visited him in hospital and ordered another scan during the hearing. This was conducted on 2nd February 2018. It revealed that nearly all of Alfie's brain had been destroyed. Fluid identical to water and cerebrospinal fluid (CSF) now replaced the damaged brain. The thalami, which controls the stimuli to the most basic sensory functions, had completely disappeared leaving Alfie with no capacity to hear, see, smell, or respond to touch except only reflexively. His brain was only able to generate seizures. He was comatose and unaware of his surroundings. From a medical perspective, continuation of care was considered futile and although it was believed that he was unlikely to suffer pain, this possibility was not ruled out altogether.

In view of these, Mr Justice Hayden held on 20th February 2018 that it was no longer in Alfie's best interests to continue to be ventilated be it at the Alder Hey Hospital or elsewhere. He made a declaration that it was lawful and in Alfie's best interests for medical treatment and all forms of ventilation to be withdrawn so as to allow him to die in dignity. He was only to be given palliative care. The trip to Italy as suggested by the parents was found not to be compatible with Alfie's best interests. Apart from being burdensome, the journey risked rendering Alfie more vulnerable including to infection, and would compromise his anticonvulsant regime.

- Court of Appeal[38]

The parents immediately sought to overturn the High Court's decision. They put forward three grounds of appeal namely that the High Court had failed to:

- Properly weigh their views in the best interests decision, and that this constituted a breach of Article 14 and Article 8 of the ECHR;

'Feared' They Would Resent New Baby" (3th September 2018), https://www.bbc.co.uk/news/uk-england-merseyside-45402094 (accessed April 7, 2019).

[38] *Re E (A Child)* [2018] EWCA Civ 550.

- Properly consider what would be an appropriate palliative care pathway and, more widely, how the consequences of the judgment were to be managed;
- Assess matters relevant to best interests or weigh up the available alternatives.

The Court of Appeal swiftly turned down grounds 2 and 3. Re ground 2, this was because the court had received reassurance from the hospital that it would not act hastily in implementing any end of life plan for Alfie. Rather, it would provide the parents with the space and time to understand any proposed palliative care plan and allow them to contribute to its development. The court likewise turned down ground 3 which alleged that Mr Justice Hayden should have given proper weight to the parents' alternative care plans when assessing Alfie's best interests. The fact that it was concluded, after hearing expert testimony and having assessed medical evidence, that it was not in Alfie's best interests to continue receiving ventilator support, meant that any alternative plans which involved the provision of a tracheostomy and gastronomy, and being transported to another country (or other countries) could not possibly be in Alfie's best interests.

The court nevertheless conceded that the parents' claim that the High Court's readiness to override parental choice in the absence of evidence of significant harm, breached Article 14 (read with Article 8) and deserved closer scrutiny. This Article states that:

> the enjoyment of the rights and freedoms set forth in this Convention shall be secured without discrimination on any ground such as sex, race, colour, language, religion, political or other opinion, national or social origin, associated with a national minority, property, birth or *other status*.

Relying on "other status," their legal team argued that it was wrong for parents to be treated differently depending on whether their application was pursued under care proceedings or the inherent jurisdiction – with the "significant harm" threshold applied to the former and "best interests" to the latter. These two proceedings were, it was claimed, appropriate comparators. Alfie's parents' situation should therefore be treated as analogous. This argument was nevertheless rejected by the Court of Appeal who opined that it was plainly wrong to put the two sets of proceedings on the same jurisdictional plane. The need to put the best interests of a sick child ahead of his parents' wishes serves as objective justification for differentiating between the two proceedings. For that reason, although permission to appeal was granted on ground 1, the appeal on that ground too was dismissed on 6th March 2018.

- Supreme Court[39]

The parents filed an application to appeal to the Supreme Court. It was their contention that it was wrong for the present issue to be determined solely by reference to the best interests test. Like the Gards, they were insistent that the matter should be considered by reference to whether their proposals for Alfie's future would cause him to suffer "significant harm." The courts have therefore, in their view, discriminated against their Article 8 right to respect for their family life, contrary to Article 14. The permission to appeal was refused by Lady Hale, Lord Kerr, and Lord Wilson.

The court confirmed that the child's "best interests" was indeed the "gold standard" for decision-making in this field and that this test was not only used in English Law but also used and supported internationally. To provide doctors with legal certainty, the court emphasised that the gold standard needed to apply to the present proceedings without qualification. It stressed that it is not lawful for doctors to provide medical treatment which is not in the child's best interests. On these bases, the Supreme Court reconfirmed on 20th March 2018 that it was not in Alfie's best interests for LST to be continued and for him to be transported by air ambulance to Italy.

- European Court of Human Rights[40]

Without delay, the parents brought the case before the ECtHR. However, both their applications for a recognition that their Articles 14 and 8 rights have been violated and for an interim measure under Rule 39 of the Rules of Court to stay the order of the domestic courts (allowing the withdrawal of LST) were rejected by the court on 23rd March 2018.

39 *In the Matter of Alfie Evans* (20th March 2018).
40 *Evans v United Kingdom* (application no. 14238/18).

- High Court[41]

The case was brought back to the High Court soon after. This was upon the application of the Alder Hey Hospital who sought direction on how the order of the 20th February 2018 should be implemented. Following this application, Mr Justice Hayden endorsed the end of life care plan outlined by the hospital, which also met the approval of Alfie's guardian. This entailed the withdrawal of artificial ventilation and other treatment at a specified date and time, and for palliative care to take place at the Alder Hey Hospital itself.

This plan was rejected by Alfie's parents. They sought a writ of *habeas corpus* to release Alfie from hospital and made an application for further medical experts to be allowed access to Alfie's medical records. They argued that Alfie was being unlawfully detained at Alder Hey since the hospital refused to allow Alfie to be discharged and for his parents to remove him. As his parents, it was submitted that they have an unfettered right to make choices and exercise their right on their son's behalf. Their Article 5 right was violated by not being allowed to do so. Mr Justice Hayden nevertheless made clear that a writ of *habeas corpus* only applies to individuals who are unlawfully detained or whose civil liberties are compromised. It had no place in the present case since Alfie's liberty could not be said to have been compromised in light of his identified best interests. Mr Justice Hayden reiterated that none of the parents' alternative solutions, namely for Alfie to be taken to Italy and subsequently to Munich, cohered with Alfie's best interests. The application was therefore rejected on 11th April 2018 as being entirely misconceived.

- Court of Appeal[42]

The parents appealed to the Court of Appeal arguing that Mr Justice Hayden was wrong to apply the best interests test to an application for a writ of *habeas corpus*. It should instead have been determined by reference to whether there was any lawful basis for his alleged "detention" in hospital. To this, the Court of Appeal reemphasised how the parents'

---

[41] *Aldey Hey Children's NHS Foundation Trust v Evans, James & Evans* [2018] EWHC 818 (Fam).

[42] *Evans & James v Alder Hey Children's NHS Foundation Trust & Alfie Evans* [2018] EWCA Civ 805.

views and rights did not take precedence over Alfie's best interests. Where these were inconsistent with one another, Alfie's right came first. The court expressly agreed with Mr Justice Hayden that the application for a writ of *habeas corpus* was misconceived. There was no basis, it held, for claiming that there was a deprivation of liberty. For it to be so, there had to be a detention, and that this must have been unlawful. Since legal precedents show that restrictions that result from the administration of medical treatment do not amount to a deprivation of liberty,[43] Alfie could not be said to have been deprived of his liberty, either with reference to Article 5 or the doctrine of *habeas corpus*. Even if there was a deprivation of liberty, it would clearly be lawful since the existing and proposed end of life plan at the Aldey Hey Hospital was in Alfie's best interests whereas the same could not be said for the proposed transfer to Italy or elsewhere. Re-endorsing the best interests test as the gold standard, the appeal was dismissed on 16th April 2018.

- Supreme Court[44]

The parents at once sought permission to appeal the Court of Appeal's ruling. The Supreme Court nevertheless stressed that inasmuch as the parents did not have the right to determine Alfie's future medical treatment, they were also not able to remove Alfie from the Alder Hey Hospital for the purpose of taking him to another country when this was not without risks and could be detrimental to him. In this respect, the court stated that English Law is no different to the ECHR or to the law of the European Union.

Their complaint at this stage was that Alfie was deprived of his liberty in contravention of Article 5 of the ECHR. However, the court was of the opinion that someone who was not able to move by virtue of interventions that were taken in intensive care to keep him alive, was not being deprived of his liberty within the meaning of the Article. It seriously questioned whether an infant incapable of staying alive without mechanical ventilation and clinically assisted nutrition and hydration as provided to Alfie, could be said to have been deprived of his liberty. Further, it had hitherto been determined that it was not in Alfie's best interests to stay at the Aldey Hey Hospital and continue to be

[43] E.g. *Nielsen v Denmark* (application number 10929/84/84); *R(Ferreira) v HM Senior Coroner for Inner South London* [2017] EWCA Civ 31.
[44] *In the Matter of Alfie Evans No. 2* (20th April 2018).

kept alive, nor to travel abroad for the same purpose. It was thereby unlawful to continue to detain him for the purpose of being kept alive artificially whether in the UK or abroad. The freedom or release that Alfie was entitled to, was release from the imposition of medical procedures which were not in his best interests. By refusing the permission to appeal on the 20th April 2018, the Supreme Court also refused any further stay of the High Court's order, thereby freeing the hospital to act in Alfie's best interests.

- European Court of Human Rights[45]

On the very same day as the Supreme Court's ruling, the parents requested for an interim measure under Rule 39 for a stay on the withdrawal of Alfie's LST. They also argued that preventing Alfie to be transferred from the Alder Hey Hospital amounted to a deprivation of liberty contrary to their Article 5 rights. Both requests were rejected by the ECtHR who ruled the parents' application inadmissible on the 23rd April 2018.[46]

Alfie was thereby disconnected from artificial ventilation at 9.45pm on Monday 23rd April 2018.

- High Court[47]

As it happened, Alfie did not die immediately and continued to breathe unaided. After being in this condition for 20 hours, his parents brought the case back to the High Court on 24th April 2018. Alfie had, by then, seemed to have been conferred Italian citizenship.[48] Owing to this, and in view of his continued ability to breathe independently, the parents asked for the February declarations to be set aside and for the court to make ancillary orders that would allow Alfie to be taken immediately to Italy. The logistics for his transportation to Italy had been planned out. They argued that the availability of an alternative that would not lead to his immediate death should now be viewed as being in Alfie's best interests. This request was dismissed by Mr Justice Hayden on the same day, pointing out that in actual fact, there had been no significant

---

45 *Evans v the United Kingdom* (application no. 18770/18).

46 Ibidem.

47 *Alder Hey Children's NHS Foundation Trust v Evans, James & Evans* [2018] EWHC 953 (Fam).

48 "Alfie Evans Granted Italian Citizenship," *Catholic Herald*, 23rd April 2018.

change in Alfie's condition. While his brain stem was alive and sustained his respiration, the substantive part of his brain had disintegrated. He therefore rejected the parents' claim that Alfie's condition was better than originally contemplated by the medical team to justify a reconsideration of the order of 20th February 2018.

- Court of Appeal[49]

An application to appeal the decision was instantaneously lodged. The parents were still emphatic that their alternative plan, which would not result in Alfie's imminent death, should be deemed as being in his best interests. Represented by separate legal teams, they each presented their own grounds of appeal. The father's appeal was predicated on events that accompanied Alfie's removal from artificial ventilation, i.e. the fact that Alfie had by then continued to breathe unaided for around 42 hours represented a change in the circumstances that justified having the order of 20th February set aside. This was turned down by the Court of Appeal, according to whom expert testimony had confirmed that a period of life post-removal of ventilation was to be expected, the length of which could not be predicted with certainty. Further, not only would the alternative plan proposed not have any prospect of success, the journey abroad would expose Alfie to further epileptic seizures and deterioration of his neurological facility.

Alfie's mother, in the meantime, appealed on the grounds that their right of free movement within the European Union to access medical services from another member state (Italy) had been violated, and that the best interests test should be eroded or adapted in such cases. However, these arguments too were turned down by the court. It was pointed out that there can be no derogation from the application of the gold standard in these cases. Hence, if it was concluded in the home state by the home court that it was in Alfie's best interests to be allowed to die, another state did not have jurisdiction to come up with a conflicting decision on what was best for the patient. Her legal team also put forward the arguments that Mr Justice Hayden erred in holding that new medical evidence was not needed and that the original hearing did not consider the likelihood of Alfie surviving the extubation process. Both of these grounds were also held to be without merit as no application

---

[49] *Evans & James v Aldey Hey Children's NHS Foundation Trust & Evans* [2018] EWCA 984 (Civ).

was made before the judge to direct any further medical evidence, and the argument that he survived the extubation was erroneous as considered previously when dealing with the father's arguments on this point. Her final ground of appeal was that, by imposing a course of action on an Italian citizen, this would amount to a criminal offence against Italian law, and could lead to extradition and prosecution of the clinical staff for the part they played in carrying out the judge's order. This ground of appeal too failed. The only relevant consideration, according to the Court of Appeal, was Alfie's best interests. The impact of that decision on healthcare professionals was not a matter which courts could take into consideration. Permission to appeal was therefore refused on 25th April 2018.

Alfie passed away 3 days later on 28th April 2018, two weeks before his second birthday.

IV. The shape of things to come?

Charlie's and Alfie's stories were followed by many in the UK and around the world.[50] This was at a rate rarely seen for medical law cases. Public opinion was, and remains, divided over who should decide their fates.[51] When reflecting on the potential implications of their medico-legal journeys, the following trends can be identified.

In the parents' determination to leave no stone unturned,[52] the courts were turned to in rapid succession particularly in the case Alfie Evans. But is the courtroom the appropriate forum for such conflicts?[53] Was it instead over-used or even misused?[54] Should mediation be de-

---

50 D. Wilkinson, J. Savulescu, "Alfie Evans and Charlie Gard – Should the Law Change?: Professionals and Families Need to Work Together," *British Medical Journal* 361 (2018): k1891, doi:10.1136/bmj.k1891.

51 G. D. Coleman, "My Name was Charlie Gard," *Ethics and Medicine* 42:11 (2017): 1-4, at 1.

52 *Alder Hey Children's NHS Foundation Trust* (n 41), para. 2.

53 L. Austin, *UK Processes for Resolution of Disagreements About the Care of Critically Ill Children* (London: Nuffield Council on Bioethics, September 2018), 16-20.

54 T. M. Pope, "Guest Editorial: Charlie Gard's Five Months in Court: Better Dispute Resolution Mechanisms for Medical Futility Disputes," *Journal of Medical Ethics* 44:7 (2018): 436-437. I. Freckleton, "Responding Better to Desperate Parents: Warnings from the Alfie Evans Saga," *Journal of Law and Medicine* 25:4 (2018): 899-918.

ployed as recommended by Mr Justice Francis in *Gard*? But how well would this alternative method of dispute resolution work for medical futility conflicts[55] bearing in mind that it has reportedly been used in *Evans*,[56] yet the case still resulted in a long-running and acrimonious legal battle?

Further, several creative arguments were put forward during the protracted trials, the most recurrent of which was the preferability or alleged superiority of the "significant harm" test over the long-established "best interests" test. Although these were ultimately defeated in the two cases, the sustainability and fairness of the best interests test continue to be challenged today not least through the Gards' proposal for a *Charlie's Law* which aims to, among other things, protect parental rights by confining judicial powers only to cases where there is a risk of significant harm to the infant.[57] However, should parents have the last word on care?[58]

Their stories also illustrated a real potential for the antagonism from the courtroom battles to spill out into, and be played out in, the public arena. That this was also able to transcend geographical boundaries and gain a momentum of its own was attributable in no small measure to the power of social media.[59] Apart from engaging in a parallel process of "trial by public opinion,"[60] parties external to the courtroom proceedings offered practical assistance including moral support, suggestions of external doctors to contact and of alternative treatments, and monetary contributions particularly through crowdfunding.[61] There

---

55 K. A. Choong, "Can 'Medical Futility' Conflicts be Mediated?" *Journal of Medical Law and Ethics* 6:1 (2018): 41-53.

56 *Re E (A Child)* (n 38), para. 113.

57 Charlie Gard Foundation, "Charlie's Law." https://www.thecharliegardfoundation.org/charlies-law/ (accessed March 31, 2019).

58 A. Caplan, K. McBride Folkers, "Charlie Gard and the Limits of Parental Authority," *Hastings Center Report* (2017): 15-16, at 15.

59 B. J. Richards, "Social Media: The Unnamed Plaintiff," *Journal of Bioethical Inquiry* 15:3 (2018): 309-312, at 309.

60 N. Hammond-Browning, "When Doctors and Parents Don't Agree: The Story of Charlie Gard," *Journal of Bioethical Inquiry* 14:4 (2017): 461-468, at 466.

61 See GoFundMe at https://www.gofundme.com/please-help-to-save-charlies-life.

were also more explicit interventions like the offers of free medical treatment abroad, flight and accommodation, and even, extraordinarily, the conferral of foreign citizenship. These resulted in the judiciary unusually commenting on events outside the courtroom in their judgments. Apart from the aforementioned comment by Mr Justice Francis in *Gard* regarding interferences from politicians and people who were not sufficiently acquainted with the medical evidence, there were at least four others articulated in the *Evans* case. In a postscript, the Court of Appeal recorded its condemnation of the clandestine manner in which Alfie was examined by a few foreign doctors, recommended by supporters, who were brought into the PICU by the parents without the permission of the court nor preceded by any consultation or approval by the treating team.[62] Apart from being in contravention of section 13 of the Children and Families Act 2014,[63] the judges observed that "a line has plainly been crossed, with such behaviour being wholly unacceptable."[64] The third time the case came before the High Court, Mr Justice Hayden expressed dismay that many videos of Alfie had been posted online by his father. Whilst recognising the desperation that drove the act, the judge remarked that these videos severely compromised Alfie's dignity and privacy.[65] The second time the case went before the Court of Appeal, the court recorded its disapproval and concern that supporters had blocked hospital staff from getting into the hospital which in turn compromised the sense of safety of staff, as well as other patients and their families. A group of supporters also marched into the PICU to the concern of staff.[66] Lord Justice McFarlane, the third time the case came before the Court of Appeal, censured the involvement of legally qualified but non-practising lawyers who provided legal advice to, and draft-

[62] *Re E (A Child)* (n 38), paragraphs 133-137.

[63] Ibidem, para. 143.

[64] Ibidem, para. 144.

[65] *Alder Hey Children's NHS Foundation Trust* (n 47), para. 2.

[66] *Evans & James* (n 42), para. 66. Similar incidents took place in the *Gard* case whereby abusive messages, including death threats, were hurled online and on the street towards staff at the treating hospital - see A. Rimmer, "Charlie Gard's Parents End Legal Fight to Keep Son Alive," *British Medical Journal* 358 (2017): j3589, doi:10.1136/bmj.j3589.

ed some of the grounds of appeal for, the parents, as introducing "a darker side to what was otherwise valuable support."[67]

Interestingly, but perhaps worryingly, these general trends may signal how medical futility conflicts involving infants would be contested in the future.

[67] *Evans & James* (n 49), para. 39.

PART I

Issues and Challenges

Chapter 1

The Vexed Question of Best Interests in Decisions Relating to Infants and Medical Futility

Jo Samanta

De Montfort University, UK

I. Introduction

Seemingly intractable disagreements between parents and doctors about the treatment of gravely ill infants are neither new, nor particularly unusual.[1] In the clinical context, proposed courses of treatment are usually discussed and agreed between the clinical team and those with parental responsibility. For critically ill infants, where withdrawal of treatment is being considered, a multidisciplinary team meeting is usually held to decide on the most appropriate course of action in light of the child's condition and prognosis. Multiple meetings may be necessary particularly where the child's condition is unstable, or where the diagnosis or prognosis is in doubt. While the aim of these meetings is partly about reaching a consensus, this cannot always be achieved and recent high profile cases demonstrate very graphically the far reaching and damaging effects of fundamental disputes about withdrawal of treatment decisions based on medical futility.[2]

---

[1] M. Brazier, "An Intractable Dispute: When Parents and Professionals Disagree," *Medical Law Review* 13:3 (2005): 412–418. This case report was published almost fifteen years ago and current case law reveals that the situation has not yet changed.

[2] Recent case illustrations include: *Great Ormond Street Hospital v Yates, Gard & Gard* [2017] EWHC 972 (Fam); *King's College Hospital NHS Foundation Trust v (1) Takesha Thomas (2) Lanre Haastrup; (3) Isaiah Haastrup* [2018] EWHC 127 (Fam); *Alder Hey Children's NHS Foundation Trust v Evans, James & Evans* [2018] EWHC 953 (Fam).

In circumstances such as these, decision-making can be very difficult and when the child is believed to be dying, these situations are all the more tragic. Decisions to withhold or withdraw life-sustaining treatment are perhaps the most emotive in medical law. Parents will naturally be devastated that their future hopes for their child will not come to pass and one can empathise with those who believe that their child's life, no matter what the cost, is worth saving.[3] In recent years, these tragic circumstances have attracted significant media attention which may be a double-edged sword. Although parents and friends will undoubtedly derive considerable moral and sometimes financial support,[4] this can also serve as an opportunity for activists interested in promoting their own agendas.[5] The negative effects on hospital Trusts and their staff should also not be underestimated.[6]

Unfortunately, infants who are critically ill are not a rare phenomenon. Thousands of children in England and Wales have life-limiting or life-threatening disease with current estimates standing in the region of 40,000.[7] In paediatric intensive care units, up to seventy percent of

[3] G. Birchley, "Deciding Together? Best Interests and Shared Decision-Making in Paediatric Intensive Care," *Health Care Analysis* 22:3 (2014): 203-222.

[4] In the case of Charlie Gard, for example, more than £1.3 million pounds was raised from well-wishers to cover his medical costs. Since Charlie's death, his parents have set up a foundation using these donations. See N. Khomami. "Charlie Gard Parents Set Up Foundation with £1.3m of Donations," *The Guardian*, 17th August 2017. https://www.theguardian.com/uk-news/2017/aug/15/charlie-gard-parents-set-up-foundation-with-donations (accessed April 5, 2019).

[5] G. Hinslif, "Alfie Evans' Parents Needed Help. The Vultures Came Instead," *The Guardian*, 26 April 2018, https://www.theguardian.com/commentisfree/2018/apr/26/alfie-evans-parents-activists (accessed March 31, 2019).

[6] See, for example, "Alfie Evans: Alder Hey Hospital Defends Staff Against Abuse," *BBC News*. 25th April 2018. https://www.bbc.co.uk/news/uk-england-merseyside-43900571 (accessed April 5, 2019).

[7] L. K. Fraser, M. Miller, R. Hain, P. Norman, J. Aldridge, P. A. McKinney, R. C. Parslow, "Rising National Prevalence of Life-Limiting Conditions in Children in England," *Pediatrics* 129:4 (2012): E923-E29.

deaths are preceded by a decision to withhold or withdraw life-sustaining treatment.[8]

Life-limiting conditions are those where there is no reasonable hope of cure and includes conditions such as neurodegenerative disease. Life-threatening conditions are those whereby curative treatment might still be an option albeit in recognition that such treatment might not be successful, for example, with certain types of cancer.[9] A key consideration for all decision-makers is the standard to be applied – namely the welfare of the child and the factors that must be taken into account to determine which course of action will be in that child's best interests.

In English law, the decision-making standard to be applied is that of the best interests of the child, as objectively assessed. In these cases, although considerable weight is given to the sanctity of life, this principle is not absolute. Despite calls for reform, more exacting approaches such as "intolerability" or incorporation of a "substantial harm" test are not part of the law.[10] The best interests approach incorporates a range of factors and considerations and all decision-makers, including the court, must attempt to balance the potentially conflicting considerations to ascertain where a child's best interests lie.

This chapter considers the standard of best interests as applied by decision-makers in the context of disputes about treatment withdrawal from infants based on the grounds of medical futility. It outlines the historical development of the case law in this area and considers recent trends. It suggests ultimately that early decisions to use alternative dispute resolution, such as mediation that can be conducted in a "safe" and confidential arena, ought to be considered and implemented where appropriate at an early stage.

[8] Royal College of Paediatrics and Child Health, *Withholding and Withdrawing Life-Sustaining Treatment in Children: A Framework for Practice* (RCPCH, London, 2004).

[9] Royal College of Paediatrics and Child Health, *National Paediatrics Diabetes Audit Report 2012-13* (London, 2014).

[10] Detailed consideration of these approaches is outside the scope of this chapter.

II. Best interests

In England and Wales, the decision-making framework for very young children is governed by the Children Act 1989 as well as a growing corpus of common law.[11] Although the Act makes no specific mention of withholding or withdrawing treatment, it provides overarching principles that must be applied in all circumstances, including serious medical decisions such as these. The Act is underpinned by a holistic approach that incorporates a range of non-exhaustive factors that must be taken into account when determining the welfare of a child.[12]

In healthcare decision-making, the child's welfare is the paramount consideration.[13] The notion of "welfare" is to be interpreted in a manner coterminous with the child's well-being and interests.[14] It embraces all that relates to a child's development as a person, as well as to his or her present and future life.[15] It focuses on medium and longer-term perspectives rather than being disproportionately influenced by transient, or more short-term considerations.[16]

In the context of decision-making for children, the overarching principle of best interests derives from Article 3(1) of the United Nations Convention on the Rights of the Child which states that "[i]n all actions concerning children, whether undertaken by public or private social welfare institutions, courts of law, administrative authorities or legislative bodies, the best interests of the child shall be a primary consideration."[17] Although best interests is also the decision-making standard for older children and adults who lack capacity, in these circumstances more subjective and nuanced considerations, such as the per-

[11] The UN Convention on the Rights of the Child (to which the UK is a signatory) defines a child as a person under eighteen years, unless the age of majority is attained earlier under applicable law.

[12] Section 1(3), Children Act 1989.

[13] Section 1, Children Act 1989.

[14] *Re B (A Minor) (Wardship: Sterilisation)* [1988] AC 199, per Lord Hailsham LC at 202.

[15] *Re G (Children)* [2012] EWCA Civ 1233, *per* Munby, LJ at para. 26.

[16] *Re O (Contact: Imposition of Conditions)* [1995] 2 FLR 124, per Sir Thomas Bingham MR at 129.

[17] Convention on the Rights of the Child (1990). Article 3(1). https://www.ohchr.org/documents/professionalinterest/crc.pdf.

son's values and beliefs, assume considerable importance.[18] For very young children whose short lives will necessarily lack any established narrative, a more objective approach is inevitable. Infants will have never possessed capacity, meaning that incorporation of these more personal elements into the decision-making calculus will be impossible. For the purposes of law, it is apparent that there is no difference between the notion of welfare and best interests for children.[19] If a dispute is referred to the court for adjudication, the child's welfare is the court's "paramount consideration."[20] Any delay in the decision-making process, or deploying the consequences of a decision will be prejudicial to the welfare of the child.[21] The court will only consider making an order where this is better than making no order at all.[22]

Satisfying the welfare principle requires the court to have regard to a range of factors[23] including (where relevant); (a) the ascertainable wishes and feelings of the child concerned; (b) his physical, emotional and educational needs; (c) the likely effect on him of any change in his circumstances; (d) his age, sex, background and any characteristics of his which the court considers relevant; (e) any harm which he has suffered or is at risk of suffering; (f) how capable each of his parents, and any other person in relation to whom the court considers the question to be relevant, is of meeting his needs; (g) the range of powers available to the court under this Act in the proceedings in question.[24] When making a decision, the court must not place parental interests before the interests of the child. While these criteria outline a range of potentially relevant factors to be considered and serve as a convenient checklist for achieving the best outcome, Tolson suggests that its wide and expansive nature allows judges to make whatever decision they believe is

[18] This aligns with section 4(6) of the Mental Capacity Act 2005 which applies to persons over the age of 16. For a useful illustration of how personal values and beliefs influence the decision-making process in adults who lack capacity, see *Aintree University Hospitals NHS Foundation v James* [2013] UKSC 67; *Briggs v Briggs* [2016] EWCOP 53; *An NHS Trust v Y* [2018] UKSC 46.

[19] *Re G (Children)* [2012] EWCA Civ 1233, per Munby LJ at para. 26.

[20] Section 1(1)(b), Children Act 1989.

[21] Section 1(2), Children Act 1989.

[22] Section 1(5), Children Act 1989.

[23] Found in section 1(4), Children Act 1989.

[24] Section 1(3), Children Act 1989.

right.[25] From this perspective, best interests might serve as a smokescreen for judicial bias, paternalism, and capricious decision-making.[26] It is therefore suggested that the welfare provisions might be better viewed as guidance, and that judicial determination ought to be accompanied by clear justification for their application to the circumstances of each case.

In the clinical arena, doctors are under no obligation to maintain a patient's life at any cost. If a treatment is futile, in that it will not achieve its therapeutic aim, or if the treatment under consideration would be overly burdensome, doctors may be entitled to withdraw or withhold it. Nevertheless, the concept of futility is not without controversy. It may, for example, be used to mask unspoken value judgements about an infant's quality of life, rather than meaning it is devoid of any therapeutic effect.[27] Any proposed intervention, including withdrawing or withholding treatment in situations of medical futility, must similarly be justified on the basis of the child's best interests. The key question for the law focuses on what is meant by best interests as justification for agreeing with, or possibly rejecting, the views of either parents or doctors. In the event of an intractable disagreement, the courts have tended to decide in favour of the clinical perspective particularly where they consider that the evidence for futility is compelling. Critics, such as Brazier, suggest that the court's approach to best interests in these circumstances may be little more than an "empty mantra" and that the courts should not be too ready to sanction a narrow medical-based perspective.[28]

---

[25] R. Tolson, "The Welfare Test and Human Rights – Where's the Beef in the Sacred Cow?" https://www.familylawweek.co.uk.

[26] J. Bridgeman, *Parental Responsibility: Young Children and Health Care* (Cambridge: Cambridge University Press, 2007), 103.

[27] There is a rich literature on this topic which is beyond the scope of this chapter. See, for example, E. D. Pellegrino, "Patient and Physician Autonomy: Conflicting Rights and Obligations in the Physician-Patient Relationship," *Journal of Contemporary Health Law and Policy* 10 (1994): 47–68; N. S. Jecker, L. J. Schneiderman, "When Families Request that 'Everything Possible' be Done," *Journal of Medicine and Philosophy* 20 (1995): 145–63; L. J. Schneiderman, N. S. Jecker, A. R. Jonsen, "Medical Futility: Its Meaning and Ethical Implications," *Annals of Internal Medicine* 112:12 (1990): 949-54.

[28] M. Brazier, "An Intractable Dispute", 415.

For decision-making purposes, parents or those with parental responsibility may consent to or refuse medical treatment, provided that these decisions are made in the child's best interests. Of course, the bottom line is that approaching decisions about the treatment of a seriously ill child from different perspectives that are based upon their diverse knowledge and experiences,[29] doctors and parents may well reach different conclusions about what is best for a child. Decisions may be challenged on this basis and the courts may impose their own, once again, potentially different views. Court adjudication in cases such as Charlie Gard,[30] Alfie Evans,[31] and Isaiah Haastrup[32] all illustrate very graphically the potentially wide gulf between perspectives on an individual child's best interests.

In some of the earlier decisions in this area, best interests was considered predominantly from a medical perspective in accordance with the *Bolam* standard,[33] namely whether that decision or practice would have been deemed proper by a responsible body of opinion skilled in that particular art, at the material time which in effect applies a particularly medicalised perspective to the area. However, more recent decisions for adults,[34] as well as for children, demonstrate a gradual shift away from this position.[35] Heywood, among others, doubts whether courts are truly ever prepared to look beyond medical interests when using their potentially wide decision-making discretion.[36] The courts could choose, for example, to view best interests in a narrow, clinical sense to justify a particular outcome whereas they could view it from a wider perspective to support a result that might align more squarely with parental or carers' views.[37] It is certainly possible to envisage circumstances where the interests of siblings might also be a factor to be

29 Ibidem, 412–418.

30 *Great Ormond Street Hospital* (n 2).

31 *Alder Hey Children's NHS Foundation Trust* (n 2).

32 *King's College Hospital NHS Foundation Trust* (n 2).

33 *Bolam v Friern Hospital Management Committee* [1957] 1 WLR 582.

34 *An NHS Trust v Y* [2018] UKSC 46.

35 *Re A (Mental Patient: Sterilisation)* [2000] 1 FLR 549, per Butler- Sloss LJ at 555.

36 R. Heywood, "Parents and Medical Professionals: Conflict, Cooperation and Best Interests," *Medical Law Review* 20:1 (2012): 29.

37 J. Herring, *Relational Autonomy and Family Law* (SpringerBriefs in Law, 2014).

considered. While flexible discretion can be inherently positive, it could, at least in theory, make predicting the outcome of cases even more difficult, causing potential problems, particularly for legal advisers. For this reason, higher standards that have to be met before court intervention are considered to be justified have been suggested including "intolerability," or the more recent "significant harm" approach. Application of the latter would mean that parental views could not be interfered with unless parental decisions place the child at a significant risk of harm.[38]

The growing body of case law demonstrates a gradual metamorphosis of the best interests standard applied in the context of seriously ill infants. The precise decision-making standard applied in these tragic cases has evolved in a manner that can be seen in chronological context. In *Re B (A Minor)*,[39] for example, the development of the more contemporary approach that was seen in *Portsmouth NHS Trust v Wyatt & Wyatt*[40] was apparent. *Re B (a minor)* concerned an infant with Down's syndrome with an intestinal obstruction which would prove fatal if left untreated. The child's parents believed sincerely that surgery would not be in her best interests. Following a balancing exercise to assess her overall best interests, the court held that corrective surgery would be lawful and in the child's best interests. Nevertheless, Lord Templeman stated that he could envisage other situations where a child's life was so full of pain and suffering and its future so uncertain that the court might permit withholding treatment. In sum, *Re B* appeared to confine legitimate instances of withholding treatment to very rare and exceptional situations of serious ill health.

This view was borne out subsequently in *Re C (A Minor) (Wardship Medical Treatment)* [1989][41] which concerned a very premature infant with severe hydrocephalus. Her prognosis was very poor and she had a range of mental and physical handicaps. The Court of Appeal found that Baby C was terminally ill and that no medical or surgical treatment would alter that actuality. Non-treatment was held to be lawful on the

[38] D. S. Diekema, "Parental Refusals of Medical Treatment: The Harm Principle as Threshold for State Intervention," *Theoretical Medicine and Bioethics* 25 (2004): 243-64.
[39] *Re B (A Minor) (Wardship: Medical Treatment)* [1981] 1 WLR 1421 (CA).
[40] [2005] 1 WLR 3995.
[41] [1989] 2 All ER 782 (CA).

basis of the paramountcy of her welfare, best interests and well-being. Emphasis was given to the fact that she was dying.[42]

The next seminal case concerned the child in *Re J (A Minor) (Wardship: Medical Treatment)* [1990].[43] The infant, who was not dying, was extremely premature with profound mental and physical disabilities. According to expert opinion, the child was blind, deaf, and likely to develop spastic quadriplegia although the ability to experience pain and discomfort was likely to be the same as for a normal baby. Lord Donaldson MR carried out a balancing exercise to assess the course to be adopted. The likely pain and suffering caused by both options were weighed against any possible commensurate benefit.[44] He spoke of the "critical equation" and the quality of the child's life, while emphasising the strong presumption in favour of maintaining life as well as the presumed view of the child. Drawing from Templeman LJ in *Re B,* Taylor LJ stated that

> I am of the view that there must be extreme cases in which the court is entitled to say: "The life which this treatment would prolong would be so cruel as to be intolerable" and that "I consider the correct approach is for the court to judge the quality of life the child would have to endure if given the treatment and decide whether in all the circumstances such a life would be so afflicted as to be intolerable to that child. I say "to that child" because the test should not be whether the life would be tolerable to the decider. The test must be whether the child in question, if capable of exercising sound judgement, would consider the life tolerable."[45]

Although the standard of "intolerability" represents a higher threshold for legitimising state intervention, compared with best interests, its leaning towards a substituted judgement is difficult to reconcile with the infant patient. On the basis of its chronological age, a very young child will have had no ability to formulate its own values and beliefs,

[42] Although the evidence suggested that she was merely incurable and severely disabled.

[43] [1991] Fam 33.

[44] *Re J (A Minor)(Wardship: Medical Treatment)* [1992] 4 All ER 614.

[45] *Re J (A Minor)* [1990] 3 All ER 930, per Taylor LJ at 945.

nor appreciate the quality of its own existence. Heywood suggests that it is for this reason that intolerability, as a decision-making standard, is too vague and unpredictable in the context of the factors to be considered under the Children Act 1989 and as such represents just one aspect of a much broader consideration of a child's best interests.[46] From this perspective, "intolerability" ought not to be used to usurp a comprehensive determination of best interests, although it might be a relevant factor to be considered as part of the overall assessment. While it can be a potentially valuable guide for determining best interests, it should not be seen as a gloss on, much less a supplementary test to ascertaining a child's best interests. [47] Where treatment would be futile, the presumption in favour of preserving life is rebutted and there is no obligation on doctors to provide futile medical treatment.[48]

In *Burke* v *GMC* [2004],[49] a case that concerned an adult rather than a child, Munby J found Taylor LJ's judgment "illuminating." While emphasising that the standard of intolerability should not be treated as a quasi-statutory yardstick, he nonetheless tied the concept of "intolerability" to the best interests standard in asserting that "(1) the assessment of best interests has to be made from the point of view or perspective of the particular patient and (2) the touchstone of best interests in this context is intolerability." In the context of decision-making for infants, Brazier suggests that if society is to give genuine respect to parental views regarding the welfare of their child, so that substantial reasons are needed to displace their (responsible) decisions, a more stringent intolerability standard might provide a better test.[50] This approach would certainly seem to give greater recognition and respect for parental (beneficent) views, particularly where a child's interests are finely balanced.

The landmark decision in *Wyatt*[51] involved an 11 month-old infant with respiratory, heart, and kidney failures. She was blind and deaf and unable to initiate voluntary action. Unanimous clinical opinion was that she was severely mentally disabled and that future treatment would

46 R. Heywood, "Parents and Medical Professionals."

47 *Portsmouth NHS Trust* (n 40), para. 24.

48 *Re J (A Minor)(Wardship: Medical Treatment)* (n 43).

49 *Burke v GMC* [2004] EWHC 1879 (Admin), para. 107.

50 M. Brazier, "An Intractable Dispute."

51 *Portsmouth NHS Trust v Wyatt & Wyatt* [2005] 1 FLR 21.

cause pain and distress. At first instance, the court declared that it would not be unlawful for the hospital to withhold artificial ventilation from Charlotte in the event that this treatment became necessary. Twice, on review, the judge affirmed his original decision despite some evidence that Charlotte's condition was improving. Her parents appealed on the basis that the correct standard to be applied in the circumstances was that of intolerability, rather than best interests, and that since Charlotte had shown some improvement, this higher test was appropriate. In rejecting this argument, the Court of Appeal stated that while intolerability was a potentially valuable guide for determining best interests, it should not be seen as a gloss on, much less a supplementary test to best interests.[52] It was held that best interests must be given a "generous interpretation" such that it went beyond clinical concerns and included the child's emotional and welfare interests. A balancing exercise was necessary in which all relevant factors had to be weighed and a convenient way of doing this was to draw up a balance sheet. This seemingly ubiquitous balancing exercise can, however, be singularly unhelpful in dispute resolution as the warring parties may well disagree about whether the life a child might have in the future would be better or worse.[53] Decision-making must be logical, transparent, and defensible, yet all parties in conflict situations will argue that their own conception of best interests is exactly that. While it was recognised that these cases were likely to be highly fact sensitive, the approach for best interests decision-making was clear: "The welfare of the child is paramount, and the judge must look at the question of the child's welfare from the assumed point of view of the patient."[54] From an ethical perspective, very young children will undoubtedly have current and future orientated interests. Although safeguarding future autonomous self-determined choice will be a vestigial interest, those of beneficence and non-maleficence will be centre-stage. Current interests will be exclusively experiential and functional and geared towards pleasure and the

[52] Ibidem, para. 24.

[53] S. Parker, "The Best Interests of the Child" in *The Best Interests of the Child: Reconciling Culture and Human Rights*, ed. P. Alson (Oxford: Clarendon Oxford, 1994), 26 – 41.

[54] *Portsmouth NHS Trust* (n 51), paragraphs 87 and 90.

avoidance of pain.[55] Future developmental interests will include agency development, opportunity interests, and those that involve human relationships.

The nebulous concept of quality of life that underpinned the court's consideration in *Wyatt* often lies at the very heart of disputes between parents and doctors and even in cases where this is not expressed directly by the parties. However, in line with best interests, quality of life is an inherently subjective notion that is notoriously hard to define. It is even harder to measure.[56] In fact, no one receives training in how to measure arbitrary criteria such as these.[57] Ultimately, on a third review, following the Court of Appeal's dismissal of the parents' appeal, the judge discharged the previous declarations on the basis that the child's circumstances had changed significantly. Throughout the entire proceedings, Charlotte's parents had cared for their daughter, with fervent commitment to preserving her life. The judge spoke eloquently of the parents' enduring commitment to Charlotte's care and accepted that they were the ones who knew her best, a point reiterated several times.[58] Despite the court's acceptance of her parents' dedication, Charlotte's clinical interests were considered to be the predominant factor for assessing her best interests. It is therefore apparent that while parents have a statutory duty to provide proper medical care for their children,[59] they can consent only to treatment that is in their child's best interests. They may not ordinarily refuse treatment that may compromise their child's welfare, as objectively assessed. Where there is doubt, the courts may substitute their own judgement for that of a parent's in a manner reflective of medical paternalism of a bygone age.[60]

---

[55] A. Buchanan, D. Brock, *Deciding For Others: The Ethics of Surrogate Decision-Making* (Cambridge: Cambridge University Press, 1998), 247.
[56] P. Boddington, T. Podpadec, "Measuring Quality of Life in Theory and in Practice," in *Bioethics: An Anthology*, eds. H. Kuhse and P. Singer (Oxford: Blackwell, 1999), 273 – 82.
[57] Mrs Nuala and Professor Jack Skarisbrick (Pro-Life campaigners) to the *Wyatt* case.
[58] *Portsmouth NHS Trust* (n 51), paragraphs 31 and 34.
[59] Section 1(1) and 1(2)(a), Children and Young Persons Act 1933.
[60] A. Bainham, "Is Anything Now Left of Parental Rights?" in *Responsible Parents and Parental Responsibility*, eds. R. Probert, S. Gilmore, J. Herring (Oxford: Hart, 2009), 23-43.

The ensuing case law reflects a slow and steady march along this now well-trammelled furrow of best interests. In *Re MB*,[61] for example, Holman J again employed a balancing exercise to ascertain the infant's best interests noting that it was the court's role and duty to exercise its own independent and objective decision-making. His worked analysis was incorporated subsequently as part of the judgment. It was considered that the child's holistic experience of living was such that the child achieved a net balance of benefit from life (which included a relationship of value with his parents as well as some pleasurable experiences from touch, sight and sound). Holman J summarised the existing case law and identified several propositions from which to construct overarching principles of judicial decision-making. His final principle was that while the views and opinions of doctors and parents should be considered carefully, parental wishes are "wholly irrelevant to consideration of the objective best interests of the child save to the extent in any given case that they may illuminate the quality and value to the child of the child- parent relationship."[62]

MB was later distinguished in the factually similar case of *Re K (A Child)* [2006].[63] While the presence of normal intelligence and awareness was believed to outweigh the burden of *MB*'s extreme physical afflictions, in *Re K* the six month-old had no comparable cognition or accretion of experiences. Sir Mark Potter held that on the evidence, there was no realistic sense that one could consign to *K* the simple pleasure of being alive other than a life of pain, distress and discomfort. It was therefore considered to be in her best interests to receive palliative care and withdraw life-sustaining treatment. Similarly, in *Re OT (A Child)*,[64] it was held that escalation of treatment would not be in the infant's best interests. The child had little awareness but nonetheless experienced considerable pain and distress caused by medical treatment.

The recent case of *Gard*,[65] concerned a baby with a rare mitochondrial disease who, according to medical opinion, had reached the stage of progression where artificial ventilation should be withdrawn and the

[61] *An NHS Trust v MB* [2006] EWHC 507 (Fam).

[62] Ibidem, para. 16.

[63] *Re K (A Child)* [2006] EWHC 1007, para. 57.

[64] *Re OT (A Child)* [2009] EWCA Civ 409 (CA).

[65] *Great Ormond Street Hospital* (n 2).

infant should be allowed to die peacefully with palliative care.[66] Following diagnosis, his parents located a Professor of Neurology in the United States who was willing to provide nucleoside therapy, an experimental treatment which offered a theoretical chance of improvement. To this end, they raised £1.3 million through a crowd-funding initiative. Although the hospital applied for ethical permission to try the experimental treatment, by the time the decision had been made Charlie's brain damage was severe and irreversible. The treatment was therefore not considered to be in Charlie's best interests. Although there is some argument that nucleoside therapy would have been less detrimental than the ensuing legal disputes, by the time the case came to trial the window of opportunity was already closed. Although one might consider that experimental treatment would have been better than death (which was the only alternative),[67] the risk of iatrogenic harm cannot be ignored. For example, in the case of Baby Andrew, where the parents did not want to extend the child's life, the list of serious iatrogenic (treatment-induced) afflictions included several months of chronic lung disease caused by mechanical ventilation. In effect, life-saving treatment resulted in prolonged episodes of decreased heart rate and lack of oxygen, painful suctioning and tube replacements, blood sampling and transfusions. Baby Andrew also developed retrolental fibroplasia, numerous infections, fractures, and finally developed seizures. In effect, he was saved only to die five months later having endured several painful and expensive months of treatment side effects.[68] A decision that focused (at least primarily) on Baby Andrew's best interests would not have allowed treatment to continue for that length of time.[69]

The next major dispute involved Isaiah Haastrup, an 11 month-old infant. The crux of the dispute was whether ventilation should be withdrawn on the basis of medical futility. Isaiah had sustained catastrophic hypoxic brain damage after being born by emergency caesarean section

---

[66] Ibidem, para. 15.

[67] D. Wilkinson, J. Savulescu, "Hard Lessons: Learning from the Charlie Gard Case," http://blog.practicalethics.ox.ac.uk/2017/07/hard-lessons-learning-from-the-charlie-gard-case/.

[68] R. Stinson, P. Stinson, "On the Death of a Baby," *Journal of Medical Ethics* 7:5 (1981), 5-18.

[69] A. Buchanan, D. Brock, *Deciding For Others*, 255.

following uterine rupture.[70] The circumstances and aftermath of his birth had led to a complete loss of trust between his parents and the hospital. After protracted arguments, as well as sincerely held parental belief in the sanctity of life, the court authorised treatment withdrawal, in Isaiah's best interests, following agreement between state and independent expert evidence of medical futility.[71]

Later that year, in *Evans*,[72] similar circumstances were to be played out, and once again on the international stage. Alfie Evans was almost two years old at the first court hearing to establish whether withdrawal of life-sustaining treatment was in his best interests. At the time of the litigation, Alfie was terminally ill as a result of an undiagnosed neurodegenerative disorder.[73] Following a protracted disagreement the hospital Trust sought a declaration that continued life-sustaining treatment was both "unkind and inhumane" and was therefore not in the child's best interests. His parents established "Alfie's Army," an online support group to campaign for the right to receive further treatment and oppose treatment withdrawal. The large protests that took place outside the hospital were accompanied by verbal abuse and threatening behaviour towards hospital staff.[74] A declaration from the High Court was granted in favour of treatment withdrawal, in Alfie's best interests, and following several unsuccessful appeals treatment was withdrawn. Although previous attempts at mediation had been unsuccessful, following treatment withdrawal Alfie's father stated publicly his wish to build bridges with the clinical team and form a care plan to ensure his son could die in comfort and dignity.

---

[70] *King's College Hospital NHS Foundation Trust* (n 2).

[71] C. Davies, "My Brain-damaged Baby Able to Respond to Me, Mother Tells Court," *The Guardian* 23rd January 2018. https://www.theguardian.com/law/2018/jan/23/brain-damaged-baby-able-to-respond-mother-court-life-support-isaiah-haastrup.

[72] *Alder Hey Children's NHS Foundation Trust* (n 2).

[73] Following his death, it was discovered that his condition was due to GABA-transaminase deficiency.

[74] B. Quinn, K. Rawlinson, "Alfie Evans: Police Issue Warning over Online Abuse of Medical Staff," *The Guardian*, 25th April 2018. https://www.theguardian.com/uk-news/2018/apr/25/alfie-evans-struggling-after-treatment-withdrawn-court-told.

In sum, the somewhat vague notion of best interests is highly subjective with a wide provenance. It has been used as a mechanism to justify withholding and withdrawing active, invasive, and intolerable treatment in circumstances of perceived medical futility.[75] Nevertheless, all of these notions are inherently subjective and these idiosyncrasies may be causative factors in parental and medical disagreements.

III. Professional guidance

Evidence-based guidance is available for doctors and decision-makers regarding infants and best interests decision-making. Such guidance is referred to increasingly by the courts[76] as well as by expert witnesses. The British Medical Association, for example, suggests that the ethical underpinnings of medicine are one and the same, and that the criteria for deciding best interests for children are the same as those for adults; including whether the child has the potential to develop awareness, the ability to interact, and the capacity for self-directed action. A key consideration is whether the child is likely to suffer severe unavoidable pain and distress.[77] The guidance emphasises that it is the value of the *treatment* which is to be assessed, and not the value of the *child*. The British Medical Association asserts that while parents will generally be the best judges of their family's interests, doctors are expected to take the lead in evaluating clinical factors. Although the ostensibly empowering statement specifies that "where there is genuine uncertainty about which treatment option would be of most clinical benefit to the child or young person, parents are usually best placed and equipped to weigh the evidence and apply it to their child's own circumstances," "the authority of parents to make decisions is not unlimited … and their authority is likely to be curtailed where the decision made would be contrary to the patient's best interests." Laurie et al, call this a "catch-22" situation in that having allocated responsibility to parents to determine best interests, that decision is then assessed further by others.[78] An

[75] An *NHS Trust v D* [2000] 2 FLR 677.

[76] *NHS Foundation Trust v AB* [2014] EWHC 1031.

[77] British Medical Association, *Withholding and Withdrawing Life-Prolonging Medical Treatment*, 3rd ed. (Oxford: Blackwell, 2016), 32.

[78] G. T. Laurie, S. H. E. Harmon, G. Porter, *Mason & McCall Smith's Law and Medical Ethics*, 10th ed. (Oxford: OUP, 2016), 531.

alternative interpretation of the guidance is that parents are granted decision-making discretion where their views concur with clinical opinion, albeit not otherwise.

The Royal College of Paediatrics and Child Health in *Making Decisions to Limit Treatment in Life-Limiting and Life-Threatening Conditions in Children: A Framework for Practice*[79] sets out the circumstances under which withholding and withdrawing life-sustaining treatment might be permissible from an ethical stance. It describes situations whereby infants should be spared inappropriately invasive treatment. It considers situations where an infant's life expectancy is limited, and where quantity of life is likely to be compromised. The guidance states that life might be regarded as intolerable when there is little or no prospect of meaningful interaction with others, or the environment, and where no reasonable person would want to lead such a life nor impose on a doctor a duty to strive actively to bring it about. Although it suggests that spastic quadriplegia with accompanying cognitive and sensory deficits might be an example of such a condition, it notes that disability is a matter of individual perception, and what is tolerable for one person may be intolerable for another.

Professional guidance, as evidence-based consensus statements that are reviewed regularly in light of new developments, can provide useful objective evidence and insights for all parties concerned. Their impartiality has potential use for early dispute resolution.

IV. Dispute resolution

Although most disputes about treatment withdrawal on the grounds of medical futility can be resolved by sensitive and empathetic discussion between clinicians and families, these attempts are not always successful. Even when health professionals recognise that attitudes are becoming fixed, early involvement of neutral third parties tends to be the exception rather than the norm and is often considered to be a last

[79] Royal College of Paediatrics and Child Health. *Making Decisions to Limit Treatment in Life-Limiting and Life-Threatening Conditions in Children: A Framework for Practice* (2015), https://www.rcpch.ac.uk/resources/making-decisions-limit-treatment-life-limiting-life-threatening-conditions-children.

resort when all else has failed.[80] In a literature review commissioned by the Nuffield Council on Bioethics, Bahtia and Austin suggest that any delay or failure to consider the need for alternative forms of dispute resolution can lead to positions becoming entrenched which reduces the likelihood that anything other than court involvement will be successful.[81] However, the adversarial nature of court adjudication can lead to irrevocable breakdown of relationships and erosion of trust between parties.

In *Gard*, Mr Justice Francis recommended that in disputes regarding medical futility that some form of mediation ought to be used where the parties could have "confidential conversations to see what common ground can be reached between them" in the belief that this would lead to "a greater understanding between the parents and the clinical team."[82] In these situations, which may well be time-sensitive, the infants at the centre of the disputes will usually continue to receive treatment that is (supposedly) against their best interests.[83] In *Gard*, for example, the Supreme Court recognised that as a result of the multiple stays of the initial High Court declaration of 11th April 2017 (that withdrawal of treatment would be lawful and in the best interests of Charlie), the hospital had been obliged to continue with treatment. While the stays of declaration meant that it was not unlawful to provide life-sustaining treatment during this time, continuation of treatment was wrong on ethical and professional grounds in that this was not in Charlie's best interests.[84] The Supreme Court recognised that the grant of even a short stay meant that the court "would in some sense be com-

---

[80] This situation was recognised in *Glass* where a disagreement ultimately ended in a fracas on a hospital ward and a total breakdown in the relationship between the boy's family and hospital doctors. See *R v Portsmouth Hospitals NHS Trust ex parte Glass* [1999] EWHC Admin 343.

[81] L. Austin, "UK Processes for Resolution of Disagreements in Care of Critically Ill Children" (2018), http://nuffieldbioethics.org/wp-content/uploads/Austin-L-2018-UK-processes-for-resolution-of-disagreements-in-care-of-critically-ill-children.pdf.

[82] *Great Ormond Street v Yates, Gard & Gard* [2017] EWHC Civ 410, *per* Justice Francis at para. 130.

[83] Ibidem, para. 3.

[84] *In The Matter of Charlie Gard (19th June 2017),* per Lady Hale at para. 15.

plicit in directing a course of action which is contrary to Charlie's best interests."[85]

A repeated criticism in *Gard*[86] was that the hospital had failed in its attempts at mediation. However, there is certainly an argument that this case was not amenable to mediation in that it involved a choice of two diametrically opposed options, rather than the possibility of a compromise.[87] Hobson's choice was either to allow an unproven therapy (which was not in Charlie's best interests at the time of the hearing) or withdraw treatment (as this was not in Charlie's best interest) and allow Charlie to die.

But is mediation the correct approach for best interests disputes about withdrawal of treatment from seriously ill infants? At its heart, mediation is a flexible process that is carried out in private. It involves the mediator, who acts as a neutral third party to assist the parties to achieve common ground with a view to negotiating a resolution. It is often considered to be an ideal form of alternative dispute resolution in that it seeks to find a "win-win" solution where both parties make concessions and seek common ground, rather than the more typical "winner takes all" approach that typifies court resolution. The primary objective of mediation is to achieve attitudinal shift by one or both parties.[88] For this reason, it has been recommended as an appropriate method for resolving healthcare disputes.[89] However, it is certainly not a panacea and it might not always work. Even if it is preferable, and more cost effective than litigation, the process may not be pleasant for either side. Randolph cautions that despite common perceptions, the need for face to face confrontation might be less comfortable for the parties concerned. The emotional intensity and inevitable confrontation of a "round table discussion" that characterises mediation can be avoided by the protection offered by the layout of the courtroom and

[85] Ibidem, para. 17.

[86] *Great Ormond Street* (n 82).

[87] E. Cave, E. Nottingham, "Who Knows Best (Interests)? The Case of Charlie Gard," *Medical Law Review* 3:1 (2018): 500–513.

[88] P. Randolph, "Spoonful of Mediation Helps the Medicine Go Down," *Journal of Medical Law and Ethics* 1 (2018): 33.

[89] See for example, M. Redfern, "Healthcare Disputes: Why Mediation is the Best Medicine," *Journal of Medical Law and Ethics* 1 (2018): 19-29.

the formal procedures and decorum of the judicial process.[90] In clinical litigation situations, even if attempts at alternative dispute resolution end in resolution, there might still be a lingering suspicion that this was premature in that all the facts and circumstances might not have been investigated properly, thereby leading to unwarranted defeat or possibly under settlement.[91]

According to the Centre for Effective Dispute Resolution, the process of mediation is "a flexible process conducted confidentially in which a neutral person actively assists the parties in working towards a negotiated agreement of a dispute or difference, with the parties in ultimate control of the decision to settle and the terms of resolution."[92]

It is this emphasis on settlement, or even agreement that can cause tensions when used for the purposes of dispute resolution in best interests decision-making. For Allen, a shift in emphasis to that of "process" rather than "outcome" means that the potential "success" of mediation becomes more apparent immediately.[93] For example, in a situation where doctors sincerely believe that continuation of life-sustaining treatment is no longer in the best interests of a terminally ill infant, whereas parents are firmly of the view that any prolongation of life is of benefit to their child, what mediation can achieve realistically is the bringing together of the parties to talk through the issues in a secure and confidential environment.

Mediation offers an environment open to exploring creative options to enable parties to reach mutually acceptable solutions. However, in the context of best interests disputes regarding withdrawal or continuation of life-sustaining treatment, it is difficult to see how the parties can realistically compromise. For parents who object to withdrawal of treatment based on religious grounds, for example, a positive decision to withdraw treatment might be a non-negotiable "red-line." At the same time, it might be against the professional conscience of clinicians to continue to provide treatment which they believe has no benefit for

---

[90] P. Randolph, "Spoonful of Mediation Helps the Medicine Go Down," 31-38.

[91] M. Redfern, "Healthcare Disputes," 28.

[92] Centre for Effective Dispute Resolution. *Model Mediation Procedure* (2018 edition), para.1 at https://www.cedr.com/about_us/modeldocs/?id=21.

[93] T. Allen, "The Extremities of Mediation and the Importance of Process," *Journal of Medical Law and Ethics* 6:1 (2018): 55-73.

the patient. This situation means that one side will have to acquiesce, reflecting the adversarial arena of the courtroom where the winner takes all.

In best interests disputes about withdrawal of treatment based on medical futility, the fundamental lack of a "middle ground" has been identified by others.[94] Choong, for example, while acknowledging the very considerable benefits of mediation, sees fundamental challenges to its routine adoption in this area: in addition to the lack of a compromise position is the unequal bargaining power of the parties concerned which, based on current law, invariably favours clinical opinion.[95]

Several different interpretations are given to the concept of mediation. For the Civil Mediation Council, the crux of the exercise is to help "both sides come to an agreement"[96] and that the mediator's role is to "help parties reach a solution to their problem" that "both parties are happy to accept."[97] Disputes regarding best interests decisions hold several unique characteristics. First, in the context of futility that the patient, due to age, will not participate in the process and second, that these are invariably time-sensitive decisions,[98] in that doctors will see continuation of life-sustaining treatment as the root cause of unnecessary pain and suffering whereas family members will see the urgency of implementing treatment, or experimental therapy. Third, in a best interests dispute, the parties will be "locked in a life or death struggle" rather than participating in an after the event litigation or dispute process.[99] Doctors might well feel that their clinical autonomy is compromised. Choong suggests that in situations such as these, root cause challenges include the absence of a middle ground and having to negotiate in the context of current law which tends to favour doctors in best interests disputes. For this reason, there is no option for "win-win."

[94] See, for example, the special edition of the *Journal of Medical Law and Ethics* on "Medical Mediation" (2018), vol. 1. In particular, see K. Choong, "Can 'Medical Futility' Conflicts be Mediated?," *Journal of Medical Law and Ethics* 1 (2018): 41-53; and P. Randolph, "Spoonful of Mediation Helps the Medicine Go Down," 31-39.

[95] Ibidem, 44.

[96] Ibidem, 53.

[97] Ibidem.

[98] T. Allen, "The Extremities of Mediation," 56.

[99] Ibidem, 59.

Whether mediation is seen as a success, or not, will therefore depend, at least to some extent, on the goals of the exercise. If avoidance of litigation is considered key, then mediation might not achieve this aim. However, if other more tangential benefits are sought, such as preservation of relationships and enhancing mutual trust and respect, then mediation might offer real benefits.[100] Other theoretical benefits include greater involvement of the parties and more control over the process of conflict resolution as well as a greater array of "settlement" options. Any agreement that is reached can also be reduced to a legally binding and enforceable agreement.

However, some disputes regarding medical futility and best interests will require referral to the court such as where there is a complete breakdown in the relationship between parents and doctors. In *R* v *Portsmouth Hospitals NHS Trust ex parte Glass* [1999],[101] Lord Woolf MR stated that where there was a grave conflict between parents and treating clinicians, the matter should be brought to the court for determination of the child's best interests. On appeal, the European Court of Human Rights[102] held that the decision to impose treatment on the child in defiance of the parents' objections constituted an interference with respect for private life under Article 8. Failure to refer such a difference to a court in circumstances that did not truly amount to an emergency amounted to a breach of that right.[103] It was further asserted that the onus was on the Trust to bring the matter before the High Court in circumstances such as these.

Although current professional guidelines and policy initiatives emphasise the "partnership of care" model between doctors and the parents of seriously ill infants, when that approach fails it is doubtful whether less authoritative bodies than courts can resolve these disputes once sides become entrenched. In *Evans*, for example, although mediation was attempted, it was unsuccessful, at least from the perspective that court intervention was still required. Mediation or alternative dispute resolution might, however, successfully narrow the issues involved.

---

[100] Ibidem, 55-73.

[101] *R v Portsmouth Hospitals NHS Trust ex parte Glass* [1999] 2 FLR 905.

[102] *Glass v* United *Kingdom* (2004) 77 BMLR 120.

[103] Ibidem, para. 81.

The fundamental issue here is that the court cannot act as mediator in that the judge will ultimately make a final and binding decision. The courts are invariably presented with limited evidence from both parties. Most of the cases discussed in this chapter were characterised by acrimony that might have been handled better if a mutually acceptable, and neutral, third party had intervened at an earlier stage. The need for reform of the decision-making processes in these conflict-laden situations is becoming recognised, and pilot projects are underway to ascertain how approaches such as these might be used in practice.[104]

Resolution of disputes requires that the views of parents are heard as well as those of the treating clinicians. Mediation, as a round table event, makes it more likely that other interested parties can also be involved compared with standard court adjudication, such as members of the wider team, nursing staff and other carers, who can spend considerable time with young children who might have spent their entire lives in hospital. For this reason, it is these members of staff who will frequently have developed rapport with the child, as well as his parents, and be able to bring valuable insights into quality of life questions as well as potential pleasure, pain, and suffering.

V. Conclusion

In sum, the somewhat vague notion of best interests is based currently on a patchwork of statute, case law, and professional guidance. Historical appraisal of landmark decisions indicates that the court's approach to treatment decisions is likely to be flexible and fact-specific. While this is commendable and appropriate, it can result in loss of predictive certainty for the parties to disputes, as well as their legal advisers.

The court has an overriding duty to exercise independent and objective judgment on the basis of all of the available evidence. Best interests are quite rightly not limited to medical interests and it is for the court to determine what the child's best interests are. For this reason, the best evidence will include the views of parents, doctors as well as other members of the care team. The voices of nurses, particularly where most of a child's life has taken place inside a hospital should be an im-

[104] See, for example, the Medical Mediation Foundation commenced in 2010 with initial funding from the Department of Health. http://www.medicalmediation.org.uk/evelina-resolution-project/.

portant decision-making factor. It is often the nurses, and sometimes the parents, who have ongoing day-to-day contact and may be better able to see and appreciate the overall picture. Nevertheless, these views and perspectives can be taken into account only to the extent that they are relevant to the child's best interests, rather than the interests or opinions of others.

Court resolution of intractable disputes carries high costs for everyone concerned and for this reason all efforts ought to be made to implement alternative means of dispute resolution wherever possible and at the very earliest stages of disagreement. The window of opportunity might be narrow. Recourse to the law will always be possible but should be a matter of last resort that is reserved for those intractable disputes where all other options have failed.

Chapter 2

Best Interests: The "Gold Standard" or a Gold Plating? Should Significant Harm be a Threshold Criterion in Paediatric Cases?

William Seagrim
University of Cardiff & 9 Park Place Chambers, UK

I. Introduction: decision-making in paediatric treatment disputes

> "The gold standard, by which most of these decisions are reached, is an assessment of his best interests."[1]
>
> Lady Hale P, Lord Kerr and Lord Wilson SCJJ, Supreme Court.

Children lack capacity by virtue of their age.[2] Proxy consent is therefore required for child medical treatment.[3] Although parents[4] naturally bear that decision-making burden, parental right is not absolute.[5] The state

---

[1] *In the Matter of Alfie Evans* (20th March 2018) (Permission Hearing before Lady Hale, Lord Kerr and Lord Wilson SCJJ), para. 14.

[2] For the purposes of this chapter, "children" does not include minors aged 16 or over, statutorily empowered to provide consent to treatment under the Family Law Reform Act 1969, section 8, or those children deemed *Gillick*-competent following *Gillick v West Norfolk and Wisbech Area Health Authority* [1986] AC 112 (HL) per Lord Fraser at 169-74.

[3] S. Pattinson, *Medical Law and Ethics,* fifth edition (London: Thomson Reuters, 2017), 159.

[4] Meaning those persons with parental responsibility under Children Act 1989, section 3.

[5] N. Allen, "Care and Treatment of Those Lacking Decision-Making Capacity," in *Principles of Medical Law*, fourth edition, ed. Judith Laing, Jean McHale (Oxford: Oxford University Press, 2017), 499.

has an interest in the discharging of parental responsibility ('PR') and the raising of children.[6] It is empowered through the court to act as final arbiter if disputes arise between parents and a treating hospital.[7] But when should parental decision-making be so usurped? Is it when an intended decision is deemed inimical to a child's welfare? Or should possible state intervention only be triggered when a higher bar is met, *i.e.* when parental decisions pose a risk[8] of significant harm? In other words, what are the acceptable boundaries of state paternalism?[9]

The law of England and Wales responds in resounding harmony. The test for medical treatment is the best interests of the child ('BI'),[10] a phrase coterminous with that child's welfare being paramount. This test applies whichever legal route is pursued to determine child treatment[11] and is now a "well laid down" principle.[12] The Supreme Court has au-

[6] J. Bridgeman, *Parental Responsibility, Young Children and Healthcare Law* (Cambridge: Cambridge University Press, 2007), 84.

[7] J. Bridgeman, "The Provision of Healthcare to Young and Dependent Children: The Principles, Concepts, and Utility of the Children Act 1989," *Medical Law Review* 25:3 (2017): 371. See also *An NHS Trust v MB (A Child represented by CAFCASS as Guardian ad Litem)* [2006] EWHC 507 (Fam), [2006] 2 FLR 319 per Holman J at para. 14.

[8] Some authors suggest "significant risk," as discussed below.

[9] Identified by Bainham and Gilmore as an acute dilemma in such cases, A. Bainham, S. Gilmore, *Children – The Modern Law*, fourth edition (Bristol: Family Law, 2013), 340.

[10] First applied in this context in *Re B (A Minor) (Wardship: Medical Treatment)* [1982] 3 FLR 117, per Dunn LJ at 123.

[11] There are three possible court applications. First, an application for a specific issue order under Children Act 1989, sections 8 and 1(1). Second, an application to the High Court for declaratory relief under its inherent jurisdiction, see *Re A (Children) (Conjoined Twins: Surgical Separation)* [2001] 2 WLR 480 per Ward LJ at 511E–12E. Third, an application for wardship, see *J v C* [1970] AC 668, per Lord Guest at 697B-H. Wardship is a manner of exercising inherent jurisdiction and suitable when issues are wider than just medical decision-making as the child remains under the protection of the court.

[12] *Great Ormond Street Hospital v Yates, Gard & Gard* [2017] EWHC 972 (Fam), per Francis J at para. 11. See also *Re A (Children) (Conjoined Twins: Surgical Separation)* (n 11), per Ward LJ at 512; *Re Z (Identification: Restrictions of Publication)* [1996] 2 WLR 88, per Sir Thomas Bingham MR at 112H–14C; *Re B (A Minor) (Wardship: Medical Treatment)* (n 10).

thoritatively declared BI to be the decision-making "gold standard."[13] However, there are commentators who contend it masks a bar set too low.[14] They submit it would be wrong to act contrary to parental wishes without establishing a risk, or significant risk, of significant harm attributable to parental choice.

Absent that threshold, do "best interests" plate in gold an unjustified state interference in the private realm of PR? A critique of the legal and ethical bases of the significant harm movement's main arguments should reveal answers. It is to a fuller exposition of those arguments that this chapter first turns.

II. The arguments: its principle proponents

A. The significant harm movement[15]

Diekema's influential article challenged the BI standard.[16] Diekema claims that "significant risk of serious harm" represents the appropriate benchmark for state intervention into parental decision-making.[17] This is the first element of an eight-staged test based on the "harm principle."[18] Significant harm does not supplant BI; it acts as an

---

[13] *In the Matter of Alfie Evans* (n 1), per Lady Hale P, Lord Kerr and Lord Wilson SCJJ at paragraphs 14 and 16.

[14] D. Diekema, "Parental Refusal of Medical Treatment: The Harm Principle as Threshold for State Intervention," *Theoretical Medicine* 25:4 (2004): 243-264; S. Elliston, *The Best Interests of the Child in Healthcare* (Abingdon: Routledge-Cavendish, 2007), 37-38 and 183; D. M. Hester, K. Lang, N. A. Garrison, D. Diekema, "Agreed: The Harm Principle Cannot Replace the Best Interest Standard ... But the Best Interest Standard Cannot Replace The Harm Principle Either," *American Journal of Bioethics* 18: 8 (2018): 38-40; D. Wilkinson, J. Savulescu, "Hard Lessons: Learning from the Charlie Gard Case," *Journal of Medical Ethics* 44:7 (2018): 438 (displaying an assumption that "serious risk of significant harm" is the correct basis for overruling parental decision-making).

[15] Referred to as "a 'harm consensus' movement" by Bester in J. Bester, "The Harm Principle Cannot Replace the Best Interests Standard: Problems With Using the Harm Principle for Medical Decision Making for Children," *The American Journal of Bioethics* 18:8 (2018): 11.

[16] Diekema, "Parental Refusal," 243-264.

[17] Ibidem, 252-253.

[18] Ibidem, 250-253. Diekema's bases were the writings of John Stuart Mill and Joel Feinberg.

enabling trigger to state intervention.[19] Other commentators seek similar thresholds. Elliston,[20] Hester et al[21] and Wilkinson and Savulescu[22] have each supported "significant risk of serious harm" thresholds. Foster defends this standard as a triage test.[23] Both Kopelman[24] and Gillon[25] have advocated formulations of a threshold.[26] Appeals to incorporate significant harm tests into law were made in the proceedings regarding both Charlie Gard ('Charlie') and Alfie Evans ('Alfie'), two children sadly at the edge of life. Those legal arguments merit consideration, starting with Charlie's case.

B. Charlie's case: a significant harm threshold … sometimes[27]

Charlie suffered from an extremely rare, inherited disorder.[28] Great Ormond Street Hospital ('GOSH') sought withdrawal of life-sustaining

---

[19] D. M. Hester, K. Lang, N. A. Garrison, D. Diekema, "Agreed," 38-40.

[20] S. Elliston, *The Best Interests of the Child in Healthcare*, 37-38 and 183. Elliston's ultimate determiner was the reasonableness of parental choice.

[21] D. M. Hester, K. Lang, N. A. Garrison, D. Diekema, "Agreed," 38-39, describing it as an "intervention principle."

[22] D. Wilkinson, J. Savulescu, "Hard Lessons: Learning from the Charlie Gard Case," 438.

[23] C. Foster, "Harm: As Indeterminate as 'Best Interests', But Useful for Triage," *Journal of Medical Ethics* 42:2 (2016): 121-122.

[24] R. McDougall, L. Notini, "Overriding Parents' Medical Decisions for their Children: A Systematic Review of Normative Literature," *Journal of Medical Ethics* 40:7 (2014): 450.

[25] Gillon suggested a starting assumption that parents should decide their children's best interests unless substantial harms or substantial injustices would result, see R. Gillon, "Why Charlie Gard's Parents Should Have Been the Decision-makers About their Son's Best Interests," *Journal of Medical Ethics* 44:7 (2018): 462.

[26] Wilkinson and Savulescu argue there is "good reason" to think a harm principle should base interference with parental decision-making, D. Wilkinson, J. Savulescu, *Ethics, Conflict and Medical Treatment for Children* (Amsterdam: Elsevier, 2019): 92-94.

[27] This chapter only addresses judgments from Charlie's proceedings that are relevant to significant harm arguments. For a wider consideration of Charlie's case, see the Introductory chapter of this book.

[28] Infantile onset encephalomyopathic mitochondrial DNA depletion syndrome, "MDDS," specifically the RRM2B-related form, *Great Ormond Street*

treatment,[29] supported by Charlie's representative.[30] Charlie's parents sought his transport to the U.S. for nucleoside therapy.[31] Mr Justice Francis granted GOSH's application, finding the experimental treatment's prospects as being "as close to zero as makes no difference [...] futile"[32] and that Charlie's awful quality of life should not be sustained without hope of improvement.[33]

Charlie's parents first raised their significant harm argument on appeal.[34] Douglas described it as "bold."[35] Originally pleaded to the effect that a court should not interfere with parental medical treatment decisions save where it risked significant harm,[36] that argument was refined orally. They sought distinct categorisation of cases: "category 1," where BI should apply, and "category 2," where parental preference should only be overridden if it was likely to cause significant harm.[37] The Court of Appeal ('the CA') understood the sifting factor into "category 2" to be the existence of a "viable treatment option."[38] The CA dismissed any legal basis for that argument and determined categorisation was a non-issue in any event because Charlie had no viable treatment op-

Hospital (n 12), per Francis J at para. 52. The syndrome is helpfully explained in Wilkinson and Savulescu, *Ethics*, 4-5.

29 *Great Ormond Street Hospital* (n 12), per Francis J at paragraphs 5 and 27. GOSH applied under section 8 of the Children Act 1989 and the inherent jurisdiction. See *In the Matter of Charlie Gard* (Supreme Court, 8th June 2017), per Lady Hale at para. 6.

30 Represented by a children's guardian appointed by the Children and Family Court Advisory and Support Service (CAFCASS) (High Court Team).

31 *Great Ormond Street Hospital* (n 12), per Francis J at paragraphs 9 and 29.

32 Ibidem, para. 119.

33 Ibidem, para. 126.

34 McFarlane LJ described the submission as "a wholly new point of law," *Yates & Gard v Great Ormond Street Hospital* [2017] EWCA Civ 410, para. 54.

35 G. Douglas, "Medical Treatment – Inherent Jurisdiction – *Yates and Another v Great Ormond Street Hospital for Children NHS Foundation Trust and Another [2017] EWCA Civ 410*," *Family Law* (2017): 958.

36 *Yates & Gard* (n 34), per McFarlane LJ at para. 54.

37 Ibidem, paragraphs 56-71.

38 Ibidem, paragraphs 36 and 56-59.

tion.[39] Even if he had, Charlie was likely to suffer significant harm whilst attempting treatment.[40]

Before the Supreme Court ('the SC'), the parents confirmed that their significant harm test acted as a threshold to what was justiciable and only applied to "category 2." They argued that the CA had misunderstood; viability of treatment was not the categorising factor. "Category 2" was apparently reserved for those cases where "parents had been able to make a decision."[41] Permission was refused[42] and the European Court of Human Rights ruled an attempted appeal "manifestly ill-founded."[43]

C. *Alfie's case: significant harm, discrimination, and human rights*[44]

Alfie's parents also challenged the boundaries of state intervention but they approached it from an entirely new angle. Alfie suffered from an undiagnosed rapidly progressive disease that devastated his brain.[45] Alder Hey Children's Hospital sought withdrawal of treatment, supported by Alfie's guardian.[46] Alfie's parents sought his travel to and

---

[39] Ibidem, para. 49. The first instance factual finding of futility was not challenged on appeal and described by the CA as being of the "utmost importance."

[40] Ibidem, paragraphs 114-115.

[41] Supreme Court, "Permission to Appeal Hearing In the Matter of Charlie Gard," https://www.supremecourt.uk/news/permission-to-appeal-hearing-in-the-matter-of-charlie-gard.html (accessed September 30, 2018). This somewhat nebulous concept was never comprehensibly explained during the permission hearing.

[42] *In the Matter of Charlie Gard* (n 29), per Lady Hale at para. 11.

[43] *Gard & Others v United Kingdom* (application no. 39793/17), para. 125. The Court declared the application inadmissible.

[44] This chapter concentrates only on those decisions from Alfie's proceedings that are relevant to significant harm arguments. For a wider consideration of Alfie's case, see the Introductory chapter of this book.

[45] *Alder Hey Children's NHS Foundation Trust v Evans James & Evans* [2018] EWHC 308 (Fam), per Hayden J at paragraphs 4-17, 29-30 and 57; *Re E (A Child)* [2018] EWCA Civ 550, per King LJ at para 13.

[46] *Alder Hey Children's NHS Foundation Trust* (n 45), per Hayden J at paragraphs 1 and 54.

treatment at Ospedale Pediatrico Bambino Gesù, Rome.[47] Despite declaring that life had intrinsic value and therefore futility did not mean axiomatically withdrawal of ventilation, in a compelling judgment, Mr Justice Hayden granted the NHS Trust's application.[48]

Alfie's parents also raised their significant harm argument for the first time in the CA. They alleged breaches of human rights based upon discrimination.[49] They claimed that parents facing the state removal of children were comparable to parents facing orders regarding disputed paediatric treatment. They further claimed that the latter were unjustifiably treated differently because state-sanctioned child removal requires satisfaction of a significant harm threshold[50] and court-imposed medical treatment decisions do not.[51] The CA flatly rejected any discrimination and principally found the analogy erroneous.[52] The SC described the comparison as "the nub" of the argument and asked itself,

> If significant harm (or its likelihood) has to be established before a child can be removed – perhaps only temporarily – from the home of his parents under a care order, why does it not need to be established before he can be removed, permanently, from them and from everything in this world, by death?[53]

The SC decided that a "powerful extra objective" was in play in care proceedings and was the differentiating factor, namely the avoidance of

---

[47] Ibidem, paragraphs 37-39. If treatment failed, the parents were possibly seeking treatment in Munich and thereafter for Alfie to die at home.

[48] Ibidem, paragraphs 51 and 64-66.

[49] *Re E (A Child)* (n 45), per King LJ at paragraphs 62-75. The parents argued a breach of the European Convention on Human Rights, Article 14 (the right against discrimination, coupled with Article 8 (the right to respect of private and family life) [due to Article 14 being parasitic of other rights]), and a breach of Article 8.

[50] Children Act 1989, section 31. Any order placing a child in the care, or interim care, of a local authority would in fact be made in accordance with BI under Children Act 1989, section 1.

[51] Although the parents' argument concentrated on the inherent jurisdiction of the High Court, a threshold would not be required under section 8 of the Children Act 1989 either.

[52] *Re E (A Child)* (n 45), per King LJ at paragraphs 80-111.

[53] *In the Matter of Alfie Evans* (n 1), para.12.

social engineering.[54] It refused permission to appeal[55] and an attempted appeal to Strasbourg was also dismissed.[56]

III. A (not so) solid legal foundation?

The foregoing arguments beg a question: is there any force to the legal arguments for a significant harm threshold in child medical treatment disputes?

A. Statutory basis: conspicuous by its absence

An NHS Trust seeking court determination of disputed medical treatment, such as in Alfie and Charlie's cases, really has two options. One is an order under the High Court's inherent jurisdiction, originating from the Crown's protective prerogative power as *parens patriae*.[57] The other is a specific issue order (SIO) pursuant to the Children Act 1989.[58] Although significant harm supporters draw analogy to forced state removal in public children law, the latter schema's significant harm threshold is a creature entirely of statute and confined to use by local authorities[59] based upon substandard parenting or a child being beyond parental control.[60] Unequivocally, there is no legislative provision for parents to pray in aid a significant harm threshold to prevent state interference with parental medical decision-making. That was a legislative choice.

---

[54] Ibidem, paragraphs 15-16.
[55] Ibidem, para. 18.
[56] *Evans v United Kingdom* (application no. 14238/18).
[57] Defined as the state in its capacity as provider of protection to those unable to care for themselves, B. Garner, ed., *Black's Law Dictionary,* tenth edition (Minnesota: Thomson Reuters, 2014): 1287; N. Allen, "Care and Treatment," 485, 494-5 and 497.
[58] Children Act 1989, section 8. NHS Trusts require permission to apply for section 8 orders, under Children Act 1989, section 10. In practice, permission is unproblematic. An NHS Trust should normally apply for both a SIO and declaratory relief, *Re JM (A Child)* [2015] EWHC 2832 (Fam), [2016] 2 FLR 235, per Mostyn J, paragraphs 20-28.
[59] Children Act 1989, section 31(1). An NHS Trust will not fall within the narrow definition of "authorised person" under section 31(9).
[60] Ibidem, section 31(2).

B. A common law basis: is a threshold justified under jurisprudence?

Does the rich common law of England and Wales save significant harm supporters? In the CA, Charlie's parents attempted to argue that a previous High Court authority, *Re King*,[61] identified the potential application of a significant harm threshold to some medical treatment cases.[62] Charlie's parents have company. Coulson-Smith et al, whilst defending BI, import a "significant harm" threshold into decision-making based upon *Re King*.[63] Is their contention sustainable?

Re King involved a five-year-old boy, Ashya, who suffered from brain cancer and was being treated by an NHS Trust. His parents, who disputed the hospital's proposed treatment plan and sought alternative treatment in Prague, took him abroad without hospital agreement.[64] The local authority was concerned Ashya was at risk of significant harm and out of the country. They applied for wardship.[65] By the time of Mr Justice Baker's first instance decision, all medical treatment issues were uncontested.[66] The judgment is therefore, understandably, a mainly narrative historical account. Within a short summary of the law, the following passage appears (later used by Charlie's parents to base their significant harm argument):

> In most cases, the parents are the best people to make decisions about a child and the State – whether it be the court, or any other public authority – has no business interfering

[61] *Re King* [2014] EWHC 2964 (Fam), [2014] 2 FLR 855.

[62] *Yates* (n 34), per McFarlane LJ, paragraphs 54 and 62-63.

[63] P. Coulson-Smith, A. Fenwick, A. Lucassen, "In Defense of Best Interests: When Parents and Clinicians Disagree," *American Journal of Bioethics* 18:8 (2018): 68.

[64] Ashya suffered from a medulloblastoma in the area of the cerebellum and posterior fossa. He had undergone brain surgery, which resected the tumour. The hospital proposed chemotherapy and radiotherapy and the parents sought a type of radiotherapy called proton therapy. *Re King* (n 61), per Baker J at paragraphs 3-11.

[65] Following a lawful removal of a child abroad and lawful retention there, if the Hague Convention does not apply, the only basis upon which a child can be returned to the jurisdiction of England and Wales is through wardship, *Re S (Wardship: Peremptory Return)* [2010] EWCA Civ 465, 2 FLR 1960, per Wall LJ at para. 14.

[66] *Re King* (n 61), per Baker J at paragraphs 20-26.

> with the exercise of parental responsibility unless the child is suffering or is likely to suffer significant harm as a result of the care given to the child not being what it would be reasonable to expect a parent to give.[67]

Bridgeman described this as a "tenuous" basis for Charlie's parents' argument.[68] Context is crucial to avoid misunderstanding the impact, if any, of *Re King*. Mr Justice Baker's comment was made at first instance, without legal argument, regarding a local authority's application for the exercise of wardship (not an NHS Trust's application for treatment decisions), and in circumstances where medical treatment was unopposed and might always have been.[69] The comment was of a general nature about applications in public children law (note the reference to substandard parenting)[70] and it was necessary in light of a crucial factor: the applicant local authority's aim was "to track down and apprehend the parents."[71] Notwithstanding the foregoing, as Lord Justice McFarlane outlined in Charlie's case, even if Mr Justice Baker had been applying a significant harm test to child treatment decisions,

> a one sentence statement in the course of a short judicial endorsement of a consent order where no point of law had been an issue, no authority had been cited, and where the judge makes no attempt to justify such a radical development of, or departure from, previous, long-established authority, provides the very weakest of bases for [parents' counsel's] weighty submissions.[72]

Bridgeman was evidently too kind.

---

[67] Ibidem, para. 31.

[68] J. Bridgeman, "A Threshold of Significant Harm f(or) a Viable Alternative Therapeutic Option," *Journal of Medical Ethics* 44:7 (2018): 466.

[69] *Re King* (n 61), per Baker J at paragraphs 20-26. Regarding proton therapy, it was confirmed that the NHS Trust had been unable to offer it but supported it subject to reliable arrangements, funding and safe transfer.

[70] See the words "as a result of the care given to the child not being what it would be reasonable to expect a parent to give"; J. Bridgeman, "The Provision of Healthcare," 389.

[71] *Yates* (n 34), per McFarlane LJ at paragraphs 100-102.

[72] Ibidem, para. 104.

C. Do human rights provide a lifeline to parental proxies?

In Charlie's case, the European Court of Human Rights expressly avoided determining the significant harm test point. They treated it as non-determinative in light of the domestic courts' findings of risk to Charlie.[73] Although there is a danger of that non-determination sowing the seeds of hope for significant harm supporters, there are good reasons to conclude human rights provide infertile ground for significant harm arguments to grow.

It should be remembered that the European Court of Human Rights noted the "broad consensus – including in international law – in support of the idea that in all decisions concerning children, their best interests must be paramount."[74] The Supreme Court shares that view[75] and it is established that conflict between a parent's right to respect of private and family life and a child's must be balanced fairly.[76] The European Court of Human Rights dismissed Alfie's parents' application as not even having the appearance of violating rights and freedoms.[77] Add in that private family law routinely interferes with PR, that it has done since the inception of the Children Act 1989 and that that overarching framework, in part utilised for paediatric decision cases, is undoubtedly human rights-compliant.

IV. Underlying justifications: (false) analogy, parental rights and focusing on the right interests

Setting aside the extant illegality of any significant harm threshold (the law can be fickle), this chapter now addresses whether its underlying rationales are compelling. Starting by examining analogies to public children law, this section then moves to parental rights claims and concludes with the question of whether such a focus on harm is appropriate.

---

[73] *Gard* (n 43), paragraphs 119-120.

[74] Ibidem, para. 118.

[75] *In the Matter of Alfie Evans* (n 1), para. 14.

[76] European Convention on Human Rights, Article 8; A. MacDonald, *The Rights of the Child: Law and Practice* (Bristol: Family Law, 2011), 461.

[77] *Evans* (n 56).

A. *Public children law: a good comparator?*

Some significant harm supporters contend that state removal of children is comparable to orders regarding disputed paediatric decisions.[78] Analysis of that analogy must commence with an understanding of state removal, enacted in England and Wales by a suite of orders collectively referred to as public children law.

Broadly, a local authority may apply for orders invading PR to safeguard a child, including by their removal into local authority care, under a care order.[79] That order creates corporate parenting by providing PR to a local authority.[80] Such state removal is primarily based on substandard parenting[81] and when drafting legislation it was a deliberate decision not to achieve that aim merely under BI.[82] As the Supreme Court noted in Alfie's case, "[f]amilies need protection from too ready a removal."[83] The framework's aim is therefore to limit the availability of compulsory child removal to cases that genuinely warrant it.[84] Although BI are the ultimate decider,[85] draconian options only become available upon satisfaction of a threshold:

---

[78] Charlie's parents; Alfie's parents; D. Wilkinson, J. Savulescu, "Alfie Evans and Charlie Gard – Should the Law Change?" *British Medical Journal* 361:8151 (2018): K1891 (by stating "[...] it would make legal decisions about medical treatment consistent with the standard applied to other types of decision").

[79] Children Act 1989, sections 31, 33 and 35. A local authority's plan could be for that child to be rehabilitated to parents, placed with other family members, live with foster parents or be placed for adoption, severing the parent/child relationship in law, often for the child's minority. Interim emergency measures are also available to the court, Children Act 1989, sections 38 and 44.

[80] Ibidem, section 33(3)(a). Under a care order, a local authority has the power to determine the extent to which parents may exercise their PR, section 33(3)(b).

[81] Ibidem, section 31(2).

[82] S. Cretney, *Family Law in the Twentieth Century* (Oxford: Oxford University Press, 2003), 726-727.

[83] *In the Matter of Alfie Evans* (n 1), para. 15.

[84] *Re J (Children) (Care Proceedings: Threshold Criteria)* [2013] UKSC 9, [2013] 1 AC 680, per Lady Hale, paragraphs 1-2; Nigel Lowe and Gillian Douglas, *Bromley's Family Law*, eleventh edition (Oxford: OUP, 2015), 553-557.

[85] Children Act 1989, section 1. If adoption is sought, a similar welfare-based test and checklist apply to any application for a placement order, to permit the

> A court may only make a care order or supervision order if it is satisfied
> - that the child concerned is suffering, or is likely to suffer, significant harm; and
> - that the harm, or likelihood of harm, is attributable to the care given to the child, or likely to be given to him if the order were not made, not being what it would be reasonable to expect a parent to give to him; or the child's being beyond parental control.[86]

Care proceedings actually protect children from poor parental care. Child neglect and child abuse are not comparable *per se* to the parenting in child medical treatment decision cases.[87] For example, at first instance in Charlotte Wyatt's case, Mr Justice Hedley explained that the court was not considering the care given to her but an "entirely organic" condition.[88] In a parenting sense, parents such as hers are blameless. Clinging to the last ounce of hope when faced with the unbounded tragedy of your own child's premature mortality is not falling short of good enough parenting. In Alfie's case, Lady Justice King expressed the following view:

> These unhappy parents, finding themselves the victims of an appalling twist of fate, are not the types of parents who find themselves the subject of care proceedings.[89]

The Supreme Court echoed a similar view.[90] An NHS Trust is not seeking to parent, it seeks guidance in respect of whether treatment is

---

local authority to place the child with potential adopters, and then to any adoption order, Adoption and Children Act 2002, section 1.

86 Children Act 1989, section 31(2).

87 For example, see the positive terms used to describe Charlie's parents, *Great Ormond Street Hospital* (n 12), per Francis J at paragraphs 10 and 128, and Alfie's parents, *Alder Hey Children's NHS Foundation Trust* (n 45), per Hayden J at para. 54.

88 *Re Wyatt (A Child) (Medical Treatment: Parents Consent)* [2004] EWHC 2247 (Fam), [2005] 1 FLR 21 [33], affd [2005] EWCA Civ 1181, [2005] 1 WLR 3995.

89 *Re E (A Child)* (n 45), per King LJ at para. 101.

90 *In the Matter of Alfie Evans* (n 1), para. 17.

or is not best for that patient. The aims and effects of public children law are distinguished.

B. Private children law: the true comparator

Significant harm supporters have conveniently ignored that the state routinely interferes with PR without clearing a care order's high hurdle.[91] Through the court in private family law cases[92] the state interferes with PR regularly[93] upon the application of a parent or person sufficiently connected to the child.[94] Courts make orders determining important matters of upbringing in accordance with BI,[95] never requiring "significant harm."[96] Court-sanctioned overriding of PR is the general scheme. Removal into state care has been singled out for special treatment by the creation of a threshold. Overriding parental wishes on BI alone, courts can force matters of upbringing on parents, *e.g.* making children available to see someone, live with someone, travel abroad or attend a particular school. Is medical treatment not just another important matter of child upbringing?

---

[91] During the permission hearing in Charlie's case, Lady Hale commented that the state, as a court, intervenes in circumstances without significant harm "all the time" when parents disagree, Supreme Court, "Permission to Appeal Hearing."

[92] Pursuant to orders made under Children Act 1989, section 8; *Re E (A Child)* (n 45), per King LJ at paragraphs 88-94.

[93] In the period April 2017-March 2018, CAFCASS received a total of 41,864 new private children law cases and between April 2016-March 2017 it received 40,536 new private law cases, "Private Law Data," CAFCASS website. https://www.cafcass.gov.uk/about-cafcass/research-and-date/private-law-data/ (accessed November 25, 2018).

[94] Children Act 1989 section 10; *Re A (Children) (Conjoined Twins: Surgical Separation)* (n 11), per Ward LJ at 511E–12E.

[95] Arrangements to live with or see people are known as "child arrangements orders" and orders specifying or prohibiting specific matters of parental responsibility, aside from child arrangements, are known as "specific issue orders" or "prohibited steps orders" respectively, under Children Act 1989, section 8.

[96] Ibidem, section 1.

C. *Parental rights: a dangerous deism*

In the circumstances of children such as Charlie and Alfie, some commentators also argue towards a justifying rationale of parental rights. This contention faces two forceful challenges that will be addressed in turn.

i. Parental rights: parasitic of fidelity to welfare

To understand the true value of parental authority, one must understand its basis. Parenting is not a right exercised *over* children but a set of responsibilities *to* them.[97] A "logically necessary corollary" of those duties is the power to make decisions.[98] That power is not absolute.[99] The *obiter* comments of Mr Justice Ward poignantly illustrate this *in extremis*,

> Parents may be free to become martyrs themselves, but it does not follow that they are free in identical circumstances to make martyrs of their children [...].[100]

Parental authority is clearly not exercised for parental benefit but for the benefit of the child[101] and any power is derived from a parent's ability to advance their child's BI.[102] The Children Act 1989 intentional-

---

[97] J. Bridgeman, "Provision of Healthcare," 377.

[98] R. Hain, "Voices of Moral Authority: Parents, Doctors and What Will Actually Help," *Journal of Medical Ethics* 44:7 (2018): 459.

[99] J. Herring, *Medical Law and Ethics*, seventh edition (Oxford: Oxford University Press, 2018), 192; R. Harper, *Medical Treatment and the Law: Issues of Consent, The Protection of the Vulnerable: Children and Adults Lacking Capacity*, second edition (Bristol: Family Law, 2014), 12; M. Brazier, E. Cave, *Medicine, Patients and the Law*, sixth edition (Manchester: Manchester University Press, 2016), 456; *Re C (HIV Test)* [1999] 2 FLR 1004 (F & CA) per Butler-Sloss at p. 1021.

[100] *Re E* [1993] 1 FLR 386 (F), per Ward J at p. 394, citing U.S. judge, Justice Holmes, in *Prince v Massachusetts* (1944) 321 US Reports 158.

[101] N. Allen, "Care and Treatment," 502; *Gillick* (n 2), per Lord Fraser at 171; Brazier and Cave, *Medicine*, 449; T. Engelhardt, "Beyond the Best Interests of Children: Four Views of the Family and of Foundational Disagreements Regarding Pediatric Decision Making," *Journal of Medicine and Philosophy* 35:5 (2010): 510.

[102] G. Birchley, "Charlie Gard and the Weight of Parental Rights to Seek Experimental Treatment," *Journal of Medical Ethics* 44:7 (2018): 449.

ly used the term "parental responsibility" rather than "parental right" to emphasise parenthood's main attribute and sole justification of status.[103] Deviation from the righteous path of BI renders a parent powerless. If parental views coincidentally accord with their child's welfare, they should be followed. If not, they should be overridden. Welfare is king. This position is uncontroversial in family law. When determining a child's upbringing, parental choice is never put on a dais.[104] In Charlie's case, Lord Justice McFarlane discusses parental views,

> It goes without saying that in many cases, all other things being equal, the views of the parents will be respected and are likely to be determinative [...] it is well recognised that parents in the appalling position that these and other parents can find themselves may lose their objectivity and be willing to "try anything," even if, when viewed objectively, their preferred option is not in a child's best interests [...] the sole principle is that the best interests of the child must prevail and that must apply even to cases where parents, for the best of motives, hold on to some alternative view.[105]

It is not contended that parental views are irrelevant but there is a difference between giving respect and giving weight.[106] Realistic options are rightly considered but they are not preferred simply because a particular party has proposed them.[107] Attaching weight to an argument solely because it is a parent's is both illogical and dangerous. In these sometimes-dreadful circumstances, are parents not vulnerable to making choices that serve their own psychological or emotional

---

[103] S. Cretney, *Family Law*, 724; Lord Mackay, "Joseph Jackson Memorial Lecture – Perceptions of the Children Bill and Beyond," *New Law Journal* 139:6403 (1989): 505.

[104] Arguably the most intrinsic weight given to parents' views is when they are facing a placement order or adoption order, under Adoption and Children Act 2002, sections 21(3) & 47(2) respectively. Broadly speaking, their consent is required for the court to make those orders. However, parental consent can be dispensed with if required by welfare, sections 21(3), 47(2) and 52(1)(b).

[105] *Yates* (n 34), per McFarlane LJ at para. 112.

[106] See the approach taken at first instance in *Re Wyatt (A Child) (Medical Treatment: Parents' Consent)* (n 88), per Hedley J at paragraphs 32- 34.

[107] *Yates* (n 34), per McFarlane LJ at para. 95.

needs?[108] Why should the views of parents inherently and automatically carry independent weight? By the sheerest of fluke a court would be forced to elevate options above others. Would this weight not create a presumption in favour of following parental wishes? Would this approach reframe the focus towards parental decision-making and/or its reasonableness?[109] How would that serve child welfare?

ii. Parental rights: casting the issue incorrectly

Is it even appropriate to view the argument as incorporating a parental right to choose treatment? Fenton-Glynn persuasively suggests that parental-rights arguments place too much weight on the "adult perspective" and adopt the wrong starting point by searching for the limit to assumed parental power rather than viewing it "solely from the perspective of the child."[110] It is wise to remember that the child is the subject of the decision, not the parents. Birchley's explanation of the basis of a parental rights argument typifies the problem. Birchley claims a foundation stone of proprietarianism, *i.e.* children being parental property.[111] Such views stab at the heart of modern children law. Children are not the passive property of another, but separate entities whose rights and interests can and should be protected, distinct from the interests of their parent(s).[112] Casting these cases in terms of parental rights or attaching weight to rights so-called is tantamount to sacrificing the welfare of children at the altar of parental power.

---

108 U. Schuklenk, "Bioethics Culture Wars – 2018 Edition: Alfie Evans," *Bioethics* 32:5 (2018): 271.

109 A difficulty with such an approach would be that assessments of the reasonableness or otherwise of parental views can be keenly disputed, *e.g.* do you take into account a parent's religious view? Is a religious view based on blind faith ever "reasonable"? See the disagreement between Wilkinson and Savulescu in D. Wilkinson, J. Savulescu, *Ethics*, 114.

110 C. Fenton-Glynn, "Life-sustaining Treatment and the Best Interests of the Child: *Re Charlie Gard*" *Family Law* (August 2017): 811.

111 G. Birchley, "Charlie Gard," 449 and 451.

112 A. MacDonald, *The Rights of the Child*, 25-26. See also the development of modern child care legislation in S. Cretney, *Family Law*, 719-721.

D. Focusing on harm: a treacherously narrow scope

There is another flaw weakening the roots of significant harm threshold arguments. According to Diekema, the biggest problem with BI is that it "represents the wrong standard" and a state's sights should be aimed only at decisions that place children at "significant risk of serious harm."[113] In a similar vein, Wilkinson and Savulescu rhetorically ask why a court should step in if a decision does not pose significant harm?[114] Is such focus on harm too narrow? First, there is a danger of undervaluing harm's current role in the BI standard. Harm is already a key component. Conceptually it is one of the grounding principles,[115] legally it is subsumed within welfare[116] and practically it is far from ignored.[117] Second, some authors have suggested that harm-based views present harm avoidance as a decision-maker's prime moral obligation.[118] This is problematic when one recognises that painless acts can still be against a person's interests. As Dworkin explains,

> My interests are in play in these circumstances not because of my capacity to feel pain but because of a different and more complex set of capacities: to enjoy or fail to enjoy, to form affections and emotions, to hope and expect, to suffer disappointment and frustration. Since a creature can be killed painlessly, even after it has capacity to feel pain, it is these more complex capacities, not the capacity to feel pain, that ground a creature's interests in continuing to live.[119]

The consequences of a threshold's focus on harm are also puzzling. It would create some medical treatment decisions where parents could

---

[113] D. Diekema, "Parental Refusal," 253.

[114] D. Wilkinson, J. Savulescu, *Ethics*, 92-93.

[115] J. Bester, "The Harm Principle," 14.

[116] Children Act 1989, section 1(3)(e).

[117] Alfie's case and Charlie's case are prime examples of a court's in-depth consideration of harm.

[118] E. De Clercq, K. Ruhe, "The Best Interest Standard: An Exhaustive Guide for Medical Decision Making in Pediatrics," *American Journal of Bioethics* 18:8 (2018): 69; J. Bester, "The Harm Principle," 15; Birchley raises similar arguments in G. Birchley, "Harm is All You Need? Best Interests and Disputes About Parental Decision-Making," *Journal of Medical Ethics* 42:2 (2016): 113.

[119] R. Dworkin, *Life's Dominion* (New York: Alfred A. Knopf, 1993), 17-18.

act in a way that is not in the best interests of their child,[120] *e.g.* the child could come second to their wishes. In Alfie's case, that notion was something Lady Justice King felt was "instinctively wrong."[121] Douglas described the idea of parents doing whatever they like to their child regardless of BI as "startling" and one that would have "overturned a hundred years of legal progress in recognising the rights of children."[122] GOSH have a mission: *The child first and always.*[123] Imagine its worrying rebranding in a post-threshold world: *The child first and always (only if parental decision-making risks significant harm).*

There are further puzzling consequences of focusing on harm as a gateway to decision-making. Imagine a child, Ianto, suffering from such difficulties in his brain that he is beyond experiencing pain, or at least it could not be found on the balance of probabilities.[124] Withdrawal of Ianto's life-sustaining treatment would inevitably lead to his death. Could parents of a strong pro-life disposition with the help of a similarly minded hospital maintain his life forever in such a parlous state? Ianto cannot feel pain, true, but any positives have also been removed from the scale that could be tipped towards continuing life. The court is deprived of the ability to undertake a sophisticated, nuanced analysis of Ianto's global interests because it cannot be said that Ianto can feel pain. This treacherously narrow scope would place some children outside of the protective reach of the court.

V. The consequences of a significant harm threshold

Ignoring its dubious legal and ethical scaffolding, if a significant harm threshold were enacted would it be workable? Several previously

---

[120] See Lady Hale's perceptive questioning during submissions in the permission hearing in Charlie's case in Supreme Court, "Permission to Appeal Hearing, *In the Matter of Charlie Gard.*"

[121] *Re E (A Child)* (n 45), per King LJ at para. 124.

[122] G. Douglas, "Medical Treatment – Evans and Another."

[123] GOSH, "Who We Are," https://www.gosh.nhs.uk/about-us/who-we-are (accessed November 22, 2018).

[124] Family proceedings, including applications for a specific issue order for medical treatment, treat alleged facts in a binary manner and the standard of proof required is the balance of probabilities, *Re B (Children) (Sexual Abuse: Standard of Proof)* [2008] UKHL 35, [2009] 1 AC 11, per Lord Hoffman at paragraphs 2 and 13; and per Lady Hale at paragraphs 32 and 70.

expressed problems with such a threshold can be overcome. It is important to expose the weaknesses of those arguments. However, one concern appears insurmountable and has hitherto been overlooked: namely, the inherent illogicality that a threshold would introduce into the law.

A. Surmountable concerns

i. Non-issue: a pejorative finding?

Birchley alleges it is arguable that judges see "evaluative overtones" in a conclusion that a parent is harming their child, not present when findings are expressed in BI terms.[125] The latter are argued to be "less pejorative."[126] This is a curious concern. Why is being pejorative or not even raised? If it is *not* being suggested that judges would avoid findings because they are pejorative, any incidental unpleasantness to judicial determinations is irrelevant. If it *is* being suggested that courts avoid making pejorative findings, that is a bold claim and requires persuasive evidence.[127] Practice teaches that courts do not balk at making findings of harm. In care proceedings these must be pejorative due to the legal test, *i.e.* the harm being attributable to the *care* given to the child, or likely to be given to him if the order were not made, *not being what it would be reasonable to expect a parent to give to him.*[128] It is pejorative by definition. However, medical treatment cases invariably assess the harm attached to treatment options. Parents happen to support or op-

[125] G. Birchley, "Harm is All You Need?" 113.

[126] Ibidem. See also G. Birchley, "The Harm Threshold and Parents' Obligation to Benefit Their Children," *Journal of Medical Ethics* 42:2 (2016): 123.

[127] Birchley cites three cases about Jehovah's Witnesses refusing blood products for their children wherein judges did not use the word "harm," at G. Birchley, "Harm is All You Need?" 113. However, those cases do not establish that courts avoid findings of harm. First, only three cases are cited. Second, one of them, *Birmingham Children's NHS Trust v B* [2014] EWHC 531 (Fam), is a very short unopposed judgment and of no use in identifying general approach. Third, although the other two cases do not mention the word "harm," it is obviously considered. In *An NHS Trust v B* [2014] EWHC 3486 (Fam), Moylan J discusses risk of death at paragraphs 10 and 17; and in *M Children's Hospital NHS Foundation Trust v Mr and Mrs Y* [2014] EWHC 2651 (Fam), Cobb J discusses life expectancy at paragraphs 6 and 12.

[128] Children Act 1989, section 31.

pose them. A judgement on the risks of treatment does not have to be pejorative of parenting. They are often conducted with the utmost respect and sympathy for parents, as exemplified by Lord Justice McFarlane's judgment in Charlie's case.[129] Significant harm in paediatrics would not have to be pejorative nor would it matter if it were.

ii. Non-issue: the indeterminacy of significant harm

Attacks on significant harm thresholds have targeted their indeterminacy and subjectivity. Birchley contends that harm has elements of indeterminacy[130] and can "contain complex value judgements."[131] Birchley is right but not on the basis of the "miracle baby" cases he cites. They are presented as having "strong similarities" yet different results, thereby proving significant harm's indeterminacy. But "strong similarities" do not necessarily lead to identical results. The devil is in the detail. These authorities are poor comparators because they are distinguished from each other regarding harm.[132] Better evidence of the indeterminacy of significant harm is presented by considering different views taken on the same facts, *e.g.* Wilkinson and Savulescu's disagreement about the level of harm in Charlie's case.[133]

---

[129] *Yates* (n 34), paragraphs 9-10 (outlining very positive findings about the parents and their love and dedication to Charlie), and para. 114 (finding a risk of significant harm attributable to their proposal).
[130] G. Birchley, "The Harm Threshold," 123.
[131] G. Birchley, "Harm is All You Need?" 112-113.
[132] In *London Borough of Lambeth v Mr and Mrs O* [2011] EWHC 3453 (Fam) (Parker J), referred to as *Re E* by Birchley, findings in respect of significant harm are unclear and it is therefore of no assistance. *London Borough of Haringey v Mrs E* [2004] EWHC 2580 (Fam), [2005] 2 FLR 47, concerned a risk of significant harm arising from a child's future being founded on a parental lie, *i.e.* the child being raised as a miracle baby, see Ryder J at paragraphs 83-89. *Re D (A Child) (Nigerian Fertility Clinic: Fact-Finding)* [2012] EWHC 4231, [2013] 2 FLR 1417 (HHJ Coleridge) concerned no such intended future lie and in *London Borough of Hillingdon v AO* [2014] EWHC 75 (Fam), referred to as *Re A (a Child) (Fact Finding Hearing: Biological Parents)* by Birchley, focus of harm was on past events and not how parents would raise A and whether they would persist in any miraculous birth view, see Hogg J at paragraphs 84-86.
[133] D. Wilkinson, J. Savulescu, "Hard Lessons," 438.

As Bester states, significant harm is "at least as indeterminate, ambiguous, and complex" as best interests.[134] Is this indeterminacy fatal? No: it is only wounding. The reasoning is straightforward. Neither judges nor doctors are automatons and sometimes in these awful cases it is difficult to determine whether a child suffers harm, if so its degree and whether any triggering level of "significant" is reached. Any ethical principle is "vulnerable to disagreement based on differing value judgements."[135] Legislation and/or jurisprudence could theoretically be developed, refining a definition and steering analysis.

B. An insurmountable hurdle: illogicality

If the foregoing criticisms are surmountable, putting aside any undesirability, could a threshold work in practice? It must be the case that significant harm supporters referring to risk attributable to parental choices more accurately refer to the choices of those with PR. Caring arrangements in modern society dictate that non-parent family members, such as step-parents, sometimes raise children and children may lack a relationship with a parent(s). A threshold connected to parental choices could only operate on our social and legal planes by meaning those with PR. That is how it will be interpreted. The workability of any threshold hypothesis can be tested with a thought experiment.

Picture two parents in a post-threshold legal framework disagreeing with each other in respect of medical treatment for their child, Efa. If they cannot agree, they need someone to decide. If there is no risk of significant harm, in the face of obstinate parents Efa's treatment is beyond resolution. Significant harm supporters would therefore probably agree that the legal test for these warring parents must be BI. The views of either parent A or parent B will be overridden by the state in the form of the court. Imagine that parent A has the unequivocal backing of Efa's hospital, all medical evidence supports treatment. Medical evidence is merely expert evidence assisting the court and the BI standard still applies.

---

[134] J. Bester, "The Harm Principle," 15.

[135] J. Bester, "The Best Interest Standard and Children: Clarifying a Concept and Responding to its Critics," *Journal of Medical Ethics* (September 2018), 5. doi:10.1136/medethics-2018-105036.

Now imagine the same scenario except Efa's parents are united against treatment. The hospital, convinced of treatment being best for Efa, would not be able to seek the court's guidance and determination. Their failure to establish a significant harm threshold would estop their application. Is this logical? The scenarios involve the same patient; the same options; the same medical evidence and the same orders. Why can a court readily override the views of parent B on the basis of BI when faced with parental *disharmony* but not override parent B arguing the same thing, about the same child, with the same medical evidence, when parents *agree*? What is so special about parental harmony? Why is it elevated above BI?

Further concerns arise from Efa's case. Would you need the agreement of everyone with PR to switch to imposing a threshold? Would the views of someone with PR less involved in that child's life have the same affect? How would the disagreement of a biological father without PR but with full-involvement impact on the presence or not of a threshold? Would the PR of a special guardian[136] carry different weight? What about the corporate PR of a local authority? Why does the choice of legal test turn on parental views rather than on the orders sought and/or the actual circumstances of the child? Would a threshold apply to challenging unified PR views on other matters of child upbringing? How could a significant harm threshold be conceptually justified as being confined to medical matters? Legally speaking, would it not bleed into other matters obtained under SIOs or the inherent jurisdiction?

It is submitted that a significant harm threshold could create a sea change in the relationship between the court and families and whether issues are or are not justiciable. It is the duty of those seeking a threshold to tackle these questions, ill addressed thus far. At best, the answers are troubling. At worst, they expose the systemic illogicality of introducing a threshold into paediatric treatment decisions.

VI. Best interests: plating sin with gold?

Rallying calls for significant harm thresholds such as those considered in this chapter are often paired with criticisms of the BI standard. The

---

136 A special guardianship order provides PR to a non-parent which can be exercised to the exclusion of any other person with PR, including a parent, apart from another special guardian, under Children Act 1989, section 14A-G.

most vociferous criticism is that BI provides no clear guidance to tribunals[137] and it allows subjective whim to be veneered in objectivity. Tackling these arguments closes the chapter.

A. Subjectivity in sheep's clothing

Huxtable powerfully suggests that there is "radical subjectivity in the law, albeit in the guise of objectivity, which is tailored to the 'individual' situation presented by the patient's plight."[138] It is an inescapable truth that BI contains components of subjectivity and objectivity. So too does all judicial decision-making, including any assessment of a significant harm threshold, *e.g.* there is subjectivity in determining whether X or Y amounts to a harm and/or what that level of harm is and/or whether it amounts to being "significant"? However, in every important sense, the BI standard is objective. Kopelman described BI as being subjective because it is shaped by the surrogate decision-maker's values and objective because it is "'grounded in sound logic, good information, and scientific findings."[139] Mr Justice Holman captured a truly objective element of the process,

> I am <u>not</u> deciding what decision I might make for myself if I was, hypothetically, in the situation of the patient; nor for a child of my own if in that situation; nor whether the respective decisions of the doctors on the one hand or the parents on the other are reasonable decisions.[140]

---

[137] D. Diekema, *Parental Refusal*, 246; J. Savulescu, "Debate: The Fiction of an Interest in Death? Justice for Charlie Gard," *British Medical Journal,* Blog, published on April 26, 2017, https//blogs.bmj.com/medical-ethics/2017/04/26/debate-the-fiction-of-an-interest-in-death-justice-for-charlie-gard/; R. Huxtable, *Law Ethics and Compromise at the Limits of Life To Treat or Not to Treat* (Abingdon: Routledge, 2013), 89; A. MacDonald, *The Rights of the Child*, 184; Savulescu's view in D. Wilkinson, J. Savulescu, *Ethics*, 161; E. Salter, "Deciding for a Child: A Comprehensive Analysis of the Best Interests Standard," *Theoretical Medicine and Bioethics* 33, (2012): 189-191.

[138] R. Huxtable, *Law Ethics and Compromise*, 95.

[139] L. Kopelman, "Why the Best Interests Standard is Not Self-Defeating, Too Individualistic, Unknowing, Vague or Subjective," *American Journal of Bioethics* 18:8 (2018): 35-6.

[140] *An NHS Trust v MB (A Child represented by CAFCASS as Guardian ad Litem)* [2006] EWHC 507 (Fam), [2006] 2 FLR 319, per Holman J at para. 16 iii.

The judge is not determining what they would do in a given situation. They are applying the law, exercising an "independent and objective judgment" in the child's best interests.[141] It is plainly as objective a standard as one can be.

B. Is the BI test unclear?

Consciously coupled with an alleged lack of objectivity is the claim that BI lacks clarity. Huxtable contends that BI are difficult to "know."[142] Deconstructing BI in the abstract, he is probably right. BI is hard to pin down without facts, as it is the child's circumstances that dictate which interests become the magnetic factors in decision-making. The BI test's individualism and adaptability is not only its greatest source of criticism, here lies its greatest strength. One could not improve on the words of Mr Justice Hedley to justify the absence of a definition,

> [t]he infinite variety of the human condition never ceases to surprise and it is that fact that defeats any attempt to be more precise in a definition of best interests.[143]

Huxtable's criticism also misunderstands the nature of BI as a standard. It is a general foundational principle and as such is not meant to provide "specific guidance."[144] The facts do the heavy lifting. If critics like Huxtable were correct, the BI standard would be unworkable in practice. Its regular successful use, nationally and internationally, persuasively proves its effectiveness.[145]

---

141 *Great Ormond Street Hospital* (n 12), per Francis J at paragraphs 11-12; N. Allen, "Care and Treatment," 509.

142 R. Huxtable, *Law Ethics and Compromise*, 89.

143 *Re Wyatt* (n 88), per Hedley J at para. 23. See also J. Herring, "Re B (Wardship: Medical Treatment) [1981] 'The Child Must Live': Disability, Parents and the Law," in *Landmark Cases in Medical Law*, ed. J. Herring, J. Wall (Oxford and Portland: Bloomsbury, 2017), 76. See also *Aintree University Hospitals NHS Foundation Trust v James* [2013] UKSC 67, [2014] AC 591, per Lady Hale at para. 36; A. MacDonald, *The Rights of the Child*, 185; A. Bainham, S. Gilmore, *Children – The Modern Law*, 341.

144 J. Bester, "The Best Interest Standard and Children," 5 and 7.

145 T. Pope, "The Best Interests Standard for Health Care Decision Making: Definition and Defense," *American Journal of Bioethics* 18:8 (2018): 36-37.

Regarding its familiarity, Kopelman argues that considerable information has been amassed as to how to promote best interests.[146] This is no truer than in the domain of paediatric decision-making. There is a rich seam of authority dating back to *Re B*[147] that shapes the welfare analysis.[148] One looks to these judicial pronouncements to better understand the standard and fully appreciate its calibre. From the "intellectual milestones" outlined by the Court of Appeal in *Re Wyatt (A Child) (Medical Treatment: Continuation of Order),*[149] further expanded upon by Mr Justice Holman in *An NHS Trust v MB (A Child represented by CAFCASS as Guardian ad Litem)*,[150] to the recent recapping of principle by Mr Justice MacDonald in *Kings College Hospital NHS Foundation Trust v Thomas*,[151] the sophisticated, guided BI assessment can be gleaned. It is far from rudderless and summarised as follows:

1. The judge must decide what is in the best interests of the child (an objective approach);
2. The welfare of the child is paramount;
3. The judge must look at the question from the assumed point of view of the child;
4. There is a (very) strong presumption in favour of a course of action which will prolong life, but that is not absolute and may be outweighed if for example the pleasures and quality of life are sufficiently small and the pain and suffering or other burdens of living be sufficiently great;
5. The term "best interests" is used in the widest possible sense and includes every kind of consideration capable of impacting on the decision, encompassing medical, emotional, sensory (pleasure, pain

[146] L. Kopelman, "Best Interests Standard," 35.
[147] *Re B (A Minor)* (n 10).
[148] In addition to case law, there is guidance produced by the medical profession that provides ethical frameworks for decision making, including for decisions about life-sustaining treatment, see Vic Larcher et al, "Making Decisions to Limit Treatment in Life-limiting and Life-threatening Conditions in Children: A Framework for Practice," *Archives of Disease in Childhood* 100, Suppl 2 (2015): s 1 – 26.
[149] [2005] EWCA Civ 1181, [2005] 1 WLR 3995, per Wall J at para. 87.
[150] [2006] EWHC 507 (Fam), [2006] 2 FLR 319, per Holman J at para. 16.
[151] [2018] EWHC 127 (Fam), [2018] 2 FLR 1028, per MacDonald J at para. 69.

and suffering) and instinctive (the human instinct to survive) considerations and all other welfare issues;
6. A court is not bound to follow clinical assessment;
7. Each case is fact specific; and
8. The court must conduct a balancing exercise in which all of the factors are weighed (although it is impossible to do so mathematically).

VII. Conclusion

The argument for a significant harm threshold in paediatric cases fractures under close examination. It has no legal basis and rationales that undergird appeals to amend the law are fundamentally flawed. Comparisons between child medical treatment cases and state removal are uncritical in their thinking; ignorant of the crucial distinguishing feature, *i.e.* poor parenting, and ignorant of the myriad other ways courts invade PR in private family law absent a threshold.

Putting aside the ethical difficulties of having a significant harm threshold, its introduction into child medical treatment cases is highly likely to create illogicality in family law. Supporters of significant harm arguments have inadequately, if at all, addressed its impact on that area's wider framework.

What is the value of BI, the decision-making standard employed in the absence of a threshold? The answer: BI is the most precious standard available by which to safeguard the interests of children, including some of this world's most vulnerable such as Charlie and Alfie. Allegations of indeterminacy misunderstand the nature of BI, overlook the wealth of jurisprudence guiding the welfare analysis and/or miscategorise as weakness its strongest attribute, *i.e.* its adaptability to a particular child's unique circumstances. This flexibility provides the BI test a fineness befitting its hallmark as the "gold standard."

Acknowledgements

I would like to thank Dr Julie Doughty, Mr Owen Thomas Q.C. and Miss Cadi Dewi for their comments on an earlier draft of this chapter. Any errors remain my own.

Chapter 3

Charlie's Law: Clarifying the Legal Standard to be Used in Medical Decision-Making for Children

Sarah Sargent
University of Buckingham, UK

I. Introduction

As a result of the lengthy and emotional legal battles over the fate of Charlie Gard, his parents have proposed a law to clearly establish the legal threshold that should be used to override parental decision about medical care for their child. *Charlie's Law* proposes the use of the "significant harm" threshold rather than that of the "best interests" principle.

The question behind *Charlie's Law* appears to be a simple one: in what instances, if any, should the rights of the parents to make medical treatment decisions for their child be overridden?

The seemingly straight-forward proposal of this standard, however, challenges the way in which courts have approached overriding parental decisions for their child's medical treatment. The way in which courts have decided these matters is inconsistent, and the source and limits of their legal authority is anything but clear.

This chapter analyses the proposal in *Charlie's Law* to use the "significant harm" threshold when there are disagreements between medical providers and parents over treatment for a sick child. It argues that *Charlie's Law* would provide much needed clarity and consistency as well as a transparent basis for the legal authority used in such situations.

Decisions about what medical treatment a child should have usually falls within the purview of parents. This, along with the other large and small decisions about the raising of a child, is known as "parental responsibility" and is defined in law in section 3 of the Children Act 1989 (CA 1989).

Parental rights are understood as a very fundamental right,[1] and one that should not be too easily displaced. Yet, circumstances and events may arise where displacement in whole or in part is necessary for the well-being of the child. This balancing of rights and interests is accomplished through the CA 1989 by the use of the "significant harm" threshold in section 31 in public law child cases. Section 31 is used when decisions are made in public law child cases as to whether the child should be subject to supervision or care order. Unlike the courts' resort to the best interests standard, the reduction or removal of parental responsibility through the significant harm standard requires that the court make a specific factual finding. It requires a factual finding by the court that a child is suffering, or is likely to suffer, significant harm, at which point a court's intervention in reducing or removing parental rights can occur.[2] After the "significant harm" threshold has been crossed, the court makes its decisions with the welfare of the child as its paramount concern.[3] Furthermore, parental responsibility can also be displaced in a different way: through the use of the court's inherent jurisdiction, in making the child a ward of the court. In this instance, the court is granted parental responsibility. This is addressed at section 100 of the CA 1989. When the court uses its inherent jurisdiction, it makes use of the "best interests" standard.

While the CA 1989 is clear as to what standard must be met in order to displace parental responsibility under sections 31 and 100, the same cannot be said for the use of the court's inherent jurisdiction. If there is some threshold that must be met before the court resorts to the use of the "best interests" principle, it is ill-defined and not articulated. Unlike the "significant harm" threshold with its specifically-required factual finds, in practice, the use of inherent jurisdiction appears to require no such thing. This leaves parents and children vulnerable to its use, and the nearly unbridled discretion it affords courts. There is a concern that the "best interests" principle can simply become a tautology – bringing

[1] See, for instance, *S-B (Children)* [2009] UKSC 17 and *In the Matter of J (Children)* [2013] UKSC 9 for a discussion on preventing unnecessary state interference with family ties.

[2] Children Act 1989, section 31.

[3] See for instance, *In the Matter of J (Children)* (n 1), and Children Act 1989 sections 1 and 31.

about whatever outcome the decision-maker wishes as being in the "best interests" of the child.

Is it possible for a court to displace the rights of the parents without a clear grant of parental responsibility to either a local authority under the CA 1989 or to itself through the use of its wardship jurisdiction?

The answer seemingly should be a clear no, but as discussed in this chapter, this is not what has happened in reality. Furthermore, it should be clear what powers a court is exercising when it displaces the rights of parents to make medical decisions about children. Again, as discussed in this chapter, this is not what has happened. The decisions that courts have made also beg the question as to whether medical decision-making for children should be treated separately from the other aspects of parental responsibility and have its own threshold for displacement. There is nothing in the statute which suggests such a division of parental responsibilities, but this is the approach that is taken by the courts.

Several problems arise from this complicated legal scenario. One is the need to clearly establish the legal threshold that is to be used when considering overriding a parent's decision on medical treatment for their child. Another is to understand the limits of such legal authority as well as its source. Without this understanding, it becomes nearly impossible for courts to issue consistent and clear decisions. Thus a danger of courts making up rules as they go, arises. The proposed *Charlie's Law* would resolve these problems and create a clear and consistent standard for medical decision-making when parents and medical professionals disagree. It puts forward the use of the "significant harm" threshold rather than a "best interests" threshold as the one which would be used.

The four subsequent sections of this chapter proceed as follows:

Section II provides a discussion of the proposal for the use of the significant harm threshold in *Charlie's Law*.

Section III discusses the Court of Appeal decision in Charlie Gard's case, contrasting it with a High Court decision in Ashya King's case, pointing out the inconsistency in the approaches, and highlighting the need for legal clarity.

Section IV discusses the lack of clarity in the resort by the courts to the "best interests" principle, questioning the origin of this power and identifying the lack of clarity around its use and its limitations.

Section V discusses the potential clash of parental autonomy and legal and medical paternalism in conflicts over medical treatment for children.

Finally, some concluding thoughts are offered in Section VI.

II. The proposal: *Charlie's Law*

The use of the significant harm test found in the Children Act 1989 to govern situations where parents and medical providers disagree on provision of treatment is included in the proposed *Charlie's Law*.[4]

This proposed legislation would "protect parental rights ... by restricting court involvement to cases where there is risk of significant harm."[5] *Charlie's Law* would establish that the use of the significant harm threshold, rather than the "best interests" of the child, is the legal standard to be applied when there is a dispute between parents and medical providers about the medical treatment to be provided to a child. Parental decisions would not be overridden without a court decision that a child is suffering, or is likely to suffer "significant harm" arising from those decisions.[6]

This proposal is not revolutionary. It simply incorporates already existing statutory law.[7] If parental responsibility for decision-making is to be displaced, who then holds that parental responsibility? Is it the court? The hospital?

Charlie's Law would ensure a consistent standard which requires factual findings be applied. This would provide much needed transparency and predictability. It would also remove the indeterminacy that arises with the use of the "best interests" standard, which, surely is a move that is beneficial for children. The "best interests" principle has no corollary to the factual findings required now and as proposed under *Charlie's Law* in the application of the significant harm threshold. This problematic stance and lack of clarity in current court approaches to overriding parental authority to make medical decisions for children is

[4] Charlie Gard Foundation, "Charlie's Law," https://www.thecharliegardfoundation.org/charlies-law/ (accessed December 4, 2018).

[5] Ibidem.

[6] Ibidem.

[7] See Children Act 1989, sections 31 and 100.

made obvious through a comparison of two decisions with very different uses of the law: a High Court decision on the treatment of Ashya King[8] and a subsequent Court of Appeal decision on Charlie Gard.[9]

The Court of Appeal decision on Charlie Gard notably rejects the "significant harm" threshold which had been used in the Ashya King decision, stating that the best interests principle was the correct approach.

But where does the "best interests" standard and its use come from? When should this power be used by the courts? What is its relationship with the statutory significant harm threshold? This remains a murky area of law. Courts fail to articulate the source of this supposed authority. This is problematic when considering what is at stake: potential life and death medical decisions regarding a child, and the overriding of the decisions of the parents in such circumstances.

The next section provides a discussion of these contrasting court decisions. The comparison of these two decisions demonstrates the need for a clear and consistent legal threshold, so that courts do not approach these situations in a disparate and wildly unpredictable manner.

III. Contrasting court decisions: the need for consistency and clarity

There is a lack of consistency in the choice of law when courts are confronted with resolving disagreements between parents and medical professionals over the treatment to be provided to a child. Moreover, courts do not clearly articulate why they have chosen one approach over another.

This section discusses the conflicting approaches that were taken by the High Court in the case of Aysha King, which was then rejected by the Court of Appeal in the case of Charlie Gard. It argues that the approach taken by the High Court is the legally sound one, even if this flies in the face of decades of use of the best interests standard.

Both cases deal with the issue of who should decide, and on what legal basis, parental choices for their children's treatment should be overridden. Yet, very different legal reasoning and explanations are

[8] *Re King* [2014] EWHC 2964 (Fam).

[9] *Yates & Gard v Great Ormond Street Hospital* [2017] EWCA Civ 410.

offered in the two cases. The High Court decision in *King*[10] sets out a clear line of reasoning – using the statutory "significant harm" threshold and then moving to the wardship use of the "best interests" standard. Yet, this approach is rejected by the Court of Appeal without a clear articulation of what legal authority is being used and why. Any nexus with the statutory "significant harm" threshold is rejected.

The following paragraph is an extract from the case on Ashya King, which clearly sets out the legal positions and legal authorities being used:

> [I]t is a fundamental principle of family law in this jurisdiction that responsibility for making decisions about a child rest with his parents. In most cases, the parents are the best people to make decisions about a child and the State—whether it be a court, or any other public authority—has no business interfering with the exercise of parental responsibility unless the child is suffering or is likely to suffer significant harm as a result of the care given to the child not being what it would be reasonable to expect a parent to give.[11]

Here the court is clearly making use of the statutory "significant harm" threshold. The recognition of this as the correct threshold is unequivocal. The case originated by a filing from the local authority "seeking permission to invoke the inherent jurisdiction ... making Ashya a ward of the court."[12] In its summary of the relevant legal principles, the decision notes that the welfare of the child is the paramount consideration, and the use of the significant harm threshold in displacing parental responsibility.[13]

The judgment is also very clear that the child was made a ward of the court on evidence that he was "at risk of suffering significant harm"[14] from the medical choices of the parent.[15] It was on the basis of this risk of "significant harm" that the court made the child a ward of

[10] *Re King* (n 8).
[11] Ibidem, para. 31.
[12] Ibidem, para. 14.
[13] Ibidem, paragraphs 29 and 31.
[14] Ibidem, para. 32.
[15] Ibidem, para. 15.

the court. Once the child is a ward of the court, decisions are made using the "best interests" standard.

The court decision links the use of the statutory "significant harm" threshold to determine the need to make the child a ward of the court, with the resulting use of the "best interests" standard only once the child is a ward of the court, and not before.

Yet, in contrast, this extract from the Court of Appeal decision on Charlie Gard is rejecting the *King* approach. This inconsistency demonstrates the lack of clarity in the law – as well as the urgent need to remedy it:

> It is also argued that where there is a viable, alternative treatment then, absent the court being satisfied that the carrying out of the treatment would cause the child significant harm, the parents' views must prevail, even if their proposed outcome is not in the child's best interests. That must be the submission that is made to the court in this case given the judge's findings. The judge has held to travel to America is not in Charlie's best interests, yet it is submitted that the parents' views to the contrary must prevail in the absence of significant harm.[16]

The paragraph above misstates the law found in the CA 1989. The court ignores or overlooks the requirement within the Act that the welfare of the child is the paramount consideration in any decision "with respect to the upbringing of the child"[17] to be made, and *must* be linked with the use of the "significant harm" test as a matter of statutory law. The basis for the rejection includes an incorrect statement of the statutory principles and requirements. Somehow the Court of Appeal does not recognise the required link between the welfare of the child and the use of the "significant harm" threshold in the CA 1989.

In advocating for the use of the serious harm threshold, the lawyers for Charlie Gard argued to the Court of Appeal for the creation of two categories of medical decisions. They argued that the so-called "category 1" decisions would be handled by the best interests standard, where no "viable alternative therapeutic option to put before the court" exist-

[16] *Yates* (n 9), para. 110.

[17] Children Act 1989, section 1(1) (a).

ed.[18] So-called "category 2" cases would have "a viable alternative treatment … put forward by the parents[19]" and would be decided on the statutory "significant harm" threshold.[20]

The *Gard* decision does not offer a convincing explanation of why the use of the "best interests" test is the correct one, simply offering that "law that had been established by the highest level of authority over the past two or more decades"[21] in supporting the correctness of the best interests standard. But the fact that something has been used, even considerably in the past, does not mean it is correct. It is highly possible that the reliance upon the inherent jurisdiction of the court has been entirely misplaced.

It is beyond dispute that medical decision-making is part of the exercise of parental responsibility, and the court is only asked to intervene in exceptional situations. Parental responsibility is clearly defined at section 3 of the CA 1989.[22] Yet, how is parental responsibility to be displaced? What legal threshold should apply? The statute provides clear rules through the invocation of the "significant harm" threshold at sections 31 and 100 of the CA 1989 when there is a matter that puts a child at risk.[23] How does medical decision-making fall outside of the usual rules about parental responsibility such that the inherent jurisdiction of the court is invoked? The Court of Appeal offers no clear explanation, and neither do scholars who are proponents of the use of the "best interests" standard in such situations.

In supporting the approach which was taken through the use of the "best interests" test, Cave and Nottingham argue that "[t]he legal principles … were compassionately and correctly applied"[24] in the litigation over Charlie Gard. They go on to explain the law:

---

[18] *Yates* (n 9), para. 58.

[19] Ibidem.

[20] Ibidem.

[21] Ibidem, para. 72.

[22] Children Act 1989, section 3.

[23] Children Act 1989, section 31. See also section 100 (4)(b) which requires that there must be "reasonable cause to believe that if the court's inherent jurisdiction is not exercised with respect to the child he is likely to suffer significant harm."

[24] E. Cave, E. Nottingham, "Who Knows Best (Interests)? The Case of Charlie Gard," *Medical Law Review* 26:3 (2017): 500.

> Disputes between parents and medical professionals that cannot be resolved by mediation or otherwise must be determined by the court with respect to the welfare principle. In addressing the welfare of the child, the court considers what is in the child's best interests from the child's perspective. Whilst the parental view is a relevant consideration, the court is not obliged to act in accordance with the wishes of parents.[25]

This explanation is confused. The use of the "best interests" test is not a matter for the CA 1989. Rather it stems from the use of the court's inherent jurisdiction. Nor do the courts in the *Gard* litigation make use of the welfare principle – clearly they have chosen to make use of the "best interests" standard, and not the provisions of the CA 1989.

Close et al generally argue in favour of the "best interests" approach that was taken by the courts in the *Gard* litigation.[26] They maintain that the approach using the best interests principle is "sound."[27] They nevertheless suggest refinements to the current process, believing that parents "should be able to insist on that treatment [one that is seen as futile or not in the best interests of the child by the medical professionals] being given to the infant."[28] They also note that "[b]est interests is an appropriate test, despite the need for more clarity in its application…"[29] While they support enhanced authority for parental decision-making,[30] their view does not justify the legal basis for the use of the "best interests" test, nor explain any superiority of its use over that of the "significant harm" standard.

However, alongside those provisions, special rules have been created by the courts to deal with the aspect of parental responsibility in medical decision-making. Should this portion of parental responsibility be specifically excised from the operation of the statute and treated differently than other aspects of parental responsibility? If so, on what

[25] Ibidem, 504.
[26] E. Close, L. Willmott, B White, "Charlie Gard: In Defence of Law," *Journal of Medical Ethics* 44 (2018): 476.
[27] Ibidem, 476.
[28] Ibidem, 477.
[29] Ibidem, 479.
[30] Ibidem, 479-480.

basis? These are the questions that beg answers, and which currently have no satisfactory legal response. Yet, on that murky and ill-defined basis, courts find the power to displace parental decision-making on the basis that to do so is in the "best interests" of the child.

The differing approaches taken by the *King* and *Gard* decisions present an unresolved question as to the scope of CA 1989 sections 31 and 100. Is it limited to applications by the local authority, or, as the *King* decision indicates, does it include interference with parental responsibility by any state authority?

The Court of Appeal is dismissive of any possibility that the "significant harm" threshold in the CA 1989 could be used to make decisions regarding the medical treatment for children. Is the inherent jurisdiction of the court to be triggered any time this question arises? This is the implication from the Court of Appeal:

> If, contrary to my primary reading, Mr Justice Baker did intend to state, where a parent puts forward a viable option for treatment, that the High Court only has jurisdiction to interfere with a parent's choice of that medical treatment if the child is likely to suffer significant harm as a result, then, in my view, such a statement has no foundation as a matter of law, is contrary to established authority and is therefore plainly in error.[31]

Other than a discussion of "it has always been done that way," the Court of Appeal offers no support for its position that medical decisions are somehow to be treated as a distinct category of event that does not fall under parental responsibility.

It may be then that the "established authority" of using the "best interests" standard via the court's inherent jurisdiction and excising medical decisions as somehow distinct from other forms of the exercise of parental responsibility is "plainly in error."

This calls for a further examination of the use of the inherent jurisdiction of the court and the use of a "best interests" standard.

---

[31] *Yates* (n 9), para. 105.

IV. "Best Interests" of the child: the inherent jurisdiction of the court?

The ability of the court to invoke the "best interests" standard is often presented as the somewhat murky operation of its inherent jurisdiction. There is no agreement among commentators as to what power the court is using when it resorts to the "best interests" principle – whether this is part of the necessity doctrine of inherent jurisdiction, or is the use of *parens patriae* powers; and indeed, no agreement exists as to whether *parens patriae* is part of the inherent jurisdiction or is something altogether distinct. Just how this operates alongside or in place of statutes is something that is not clear from the cases in this area, nor is the explanation or the triggering of the use of the "best interests" principle consistent from one decision to the next. Yet, it is this powerful paternalistic power that the courts resort to again and again to override parental decisions. It is the use of this power that *Charlie's Law* seeks to curb, by putting in statute the more clearly defined serious harm threshold.

The cases of Ashya King, Charlie Gard, and the recent High Court decision in *Re C*[32] do not provide consistent explanations for the use of the "best interests" principle, nor do they reach it in the same way. And yet, all of these cases deal with the same legal issue: what happens when parents and medical providers do not agree on treatment for a child? What threshold does the court use to override parental responsibility and intervene? Or is it even necessary for the courts to override parental responsibility? Could it remain intact and somehow undisturbed, even while its ability to be exercised is circumscribed by the court as to medical decision-making?

Courts are inconsistent as to whether they offer any explanation or justification of their resort to the "best interests" principle. This is demonstrated in the *Re C* decision where the High Court insists that:

> As the law is well established so there is no need in this judgment to set out a more extensive reference to authority and case law.[33]

[32] *Re C (An Infant) (Rev 1)* [2018] EWHC 2750 (Fam).
[33] Ibidem, para. 27.

But the rendition of law that is offered is a confused mixture of statutory language and "best interests" language. Within the same decision, the court relies on the decision in *Wyatt v Portsmouth NHS Trust*[34] quoting it in part:

> *The judge must decide what is in the child's best interests. In making that decision, the welfare of the child is paramount, and the judge must look at the question from the assumed point of view of the patient.*[35]

The language on the welfare of the child standard and its paramountcy are clearly those of section 1 of the CA 1989 (the same Act that contains the "significant harm" threshold). The "best interests" standard comes from the inherent jurisdiction of the court. They are not the same thing, and should not be used in such a confused and inter-changeable manner. If it is the statute that is being used, the court should be clear. If the court is having recourse to its inherent jurisdiction, it must be clear as to how and why. This mixing of language only confuses the legal standard that is being invoked.

It matters which power the court is using, as the legal thresholds are very different. But the courts themselves seem to lack clarity; such that they are unaware of the mixing of standards at the same time they declare how clear the law is.

In writing about the inherent jurisdiction when used for vulnerable adults, Hall notes the very little information or understanding by the courts about this power. Yet, despite this, courts do not hesitate to deploy it.[36]

Hall indicates that there are two sources of inherent jurisdiction. She delineates one of these as containing "core powers"[37] that are concerned with "the self-governance functions of the court, are essential to the court's identity and as such are constitutionally protected."[38] The other contains "non-core powers"[39] which derive

[34] *Wyatt v Portsmouth NHS Trust* [2005] EWCA Civ 1181.

[35] *Re C (An Infant)* (n 32), para. 26, citing *Wyatt v Portsmouth NHS Trust*, italics in original.

[36] M. Hall, "The Vulnerability Jurisdiction: Equity, *Parens Patriae*, and the Inherent Jurisdiction of the Court," *Canadian Journal of Comparative Law* 185 2:1 (2016): footnote 19, p. 192.

[37] Ibidem, 195.

[38] Ibidem.

from the necessity doctrine, where the court acts "to effect justice and equity between the parties."[40]

Hall clearly locates the court's intervention in medical cases involving vulnerable adults as making use of the "best interests" standard as arising from the necessity doctrine. She comments that:

> In both medical and non-medical settings, intervention could be justified on the basis of necessity only if in the best interests of the individual; outside of the medical context this has been interpreted as a requirement that intervention is necessary for the purpose of safeguarding the rights of an individual who, by reason of mental incapability, is incapable of doing so herself.[41]

Moreover, Hall sees the *parens patriae* power as distinct from the inherent jurisdiction of the court.[42] This distinction is important, as Hall views the necessity doctrine as one that "must retreat where the gap has been filled by legislation."[43] The continued existence of the inherent jurisdiction in the face of legislation would have to be called into question. But this view of the inherent jurisdiction and the powers being used by the court is by no means universal. A fully different view is offered by Mr Justice Munby.

In contrast to Hall, Mr Justice Munby locates the best interests use by the courts not in the doctrine of necessity.[44] He breezily explains that the use of the "best interests" standard arises from the *parens patriae* powers of the Court:

> The outcome was the rediscovery – in plain language the invention – by the family judges of a full-blown welfare based *parens patriae* jurisdiction in relation to incapacitated

[39] Ibidem, 196.
[40] Ibidem.
[41] Ibidem, 202.
[42] Ibidem, 195.
[43] Ibidem, 196.
[44] J. Munby, "Protecting the Rights of Vulnerable and Incapacitous Adults—The Roles of the Courts: An Example of Judicial Law Making," *Child and Family Law Quarterly* 26 (2014): 64-77.

> adults which is indistinguishable from the long established *parens patriae* jurisdiction in relation to children.[45]

There is thus no agreement between the two commentators as to the source of the "best interests" standard being used by the courts. Which source is at the origin of this matters, because as Hall explains, the necessity doctrine fades when a relevant statute is enacted.

It is difficult to understand the basis upon which courts resort to the "best interests" principle. Is it the necessity doctrine? Is it *parens patriae*? And if it is *parens patriae*, is this part of the court's inherent jurisdiction or something else altogether?

Despite the strong pronouncements in decisions that the "law is clear," it is anything but, and this is to the detriment of children, parents, medical professionals, and ultimately, to the credibility of the legal basis on which these difficult situations are decided.

Even in their efforts to declaim the clear state of the law, courts combine language from the statute with the language of "best interests" principle and seem unaware of this conflation, or worse, unaware of the legal uncertainty that arises from the conflation. If the courts are unaware of this mixing of language from two sources of legal power, how much faith then can be put into their statements that their use of the "best interests" principle is correct and unassailable? All of this becomes sharply focused when realising that these questions arise in life-and-death circumstances of a child – all the more a need to make sure that the court not only comprehends its authority, but also comprehends the limits of this authority.

V. Medical paternalism: how decisions are made

The environment in which treatment decisions are made for children within the medical and legal spheres needs to be understood in order to determine what legal threshold is necessary when parents and medical professionals disagree on the treatment provided for a child. The presence and the effects of medical paternalism need to be considered.

Orfali's research indicates that medical paternalism has a substantial role to play in the medical environment.[46] Further, she indicates that a

45 Ibidem, 77.

medical provider's stance on a treatment choice can be an exercise of power – rather than purely a scientifically indicated decision.[47]

Even when it appears that parents are making decisions, the ultimate control of the medical profession is omnipresent. Orfali comments that even in a setting where parental autonomy rather than medical paternalism is the model in operation that "in reality things remain strikingly under medical authority."[48]

Orfali argues that parental autonomy – the exercise of parental responsibility – when it comes to medical decision-making for a child, is already heavily circumscribed by medical paternalism. There is no need to inject more paternalism – the use of the "best interests" standard by the court, howsoever the authority arises – when the reality is that parental autonomy already involves medical paternalism since "[p]arental autonomy can only exist insofar physicians allow such autonomy."[49]

This dynamic has to be recognised in understanding the decision-making environment and considering not only what the respective roles of parents, medical professionals, and courts should be, but also in considering how that conflicts over treatment decisions come about.

Restricting the ability to override parental decision-making through the use of the "significant harm" threshold also limits the incursion of medical paternalism into parental autonomy. The requirement of specific factual findings as part of the use of the "significant harm" threshold provides some protection against the indeterminacy which is a hallmark of the "best interests principle." Within such indeterminacy, it is easy for paternalism to flourish.

VI. Conclusion

Paris et al comment on the unsatisfactory nature of the use of courts to decide difficult medical treatment decisions for children:

> Courts are, by nature, adversarial, cumbersome and costly. They lack sensitivity to clinical decisions. Furthermore, they

[46] K. Orfali, "Parental Role in Medical Decision-Making: Fact or Fiction? A Comparative Study of Ethical Dilemmas in French and American Neonatal Intensive Care Units," *Social Science and Medicine* 28 (2004): 2009-2022.

[47] Ibidem, 2015.

[48] Ibidem, 2020.

[49] Ibidem, 2014.

> are not structured to resolve medical disputes in a timely fashion. Nor, in adjudicating a particular case in controversy, does a trial judge have the time, staff, or resources to explore complex and troubling issues in the depth required to establish sound public policy.[50]

The use of the "best interests" standard, rather than the "significant harm" threshold, may be one that encourages resort to the courts when disagreements arise between parents and medical providers. It is a much easier standard to satisfy for medical professionals who wish to remove parental decision-making than the "significant harm" threshold. Knowing that establishing a case in court might be more difficult, might encourage medical professionals to work with parents in non-adversarial contexts to arrive at a mutually agreed decision. This is another aspect of *Charlie's Law* – the proposal of the use mediation as a means to avoid litigation[51] – and this works hand-in-hand with the "significant harm" threshold to avoid the courts as the place that these disagreements are to be solved.

The indeterminacy and potential tautological reasoning of the "best interests" principle is avoided through the use of the "significant harm" threshold in the CA 1989. This threshold requires that there be factual findings made as to whether the child is suffering or is likely to suffer "significant harm."[52] These required factual findings are a bulwark against the risk of indeterminacy in the use of the "best interests" principle.

There are many gains to be had through the enactment of *Charlie's Law*. The legal threshold for court involvement and for the displacement of parental responsibility in making medical treatment decisions for children will be clear. The "significant harm" threshold will be the one that is used. As it stands, there is no agreement amongst legal decisions by the courts as to whether the "best interests" standard or the significant harm threshold is to be used. Different standards were used

---

[50] J. J. Paris, J. Ahluwalia, B. M. Cummings, M. P. Moreland, D. J. Wilkinson, "The Charlie Gard Case: British and American Approaches to Court Resolution Over Medical Decisions," *Journal of Perinatology* 37 (2017): 1268, at 1270.

[51] Charlie Gard Foundation, "Charlie's Law."

[52] See *S-B (Children)*(n 1) and *In the Matter of J (Children)* (n 1).

by the High Court in the *King* case and by the Court of Appeal in the *Gard* case. This lack of clarity operates in the interests of no-one.

The "best interests" standard itself is at best a little understood principle – with no agreement amongst commentators as to what area of the court's power it arises from. Understanding the origin of this legal authority is important as that has implications for the limits on the authority. Again, the enactment of *Charlie's Law* resolves this problem, by providing a clear statement of the court's authority and the limits on that authority.

Finally, *Charlie's Law* helps to protect the parental autonomy that is already embedded in the legal notion of parental responsibility. This is important when considering the research that indicates that medical paternalism is omnipresent in the decision-making environment already. *Charlie's Law* prevents the further erosion of parental autonomy through the operation of the "significant harm" threshold instead of the "best interests" principle.

Providing clear and transparent legal principles that are well understood by parents, medical providers and courts can only be a positive change. It is this that *Charlie's Law* would provide. It is this which is sorely needed.

Chapter 4

Media Framing of "Medical Futility": Flaming the Debate?

Kim McGuire
University of Central Lancashire, UK

I. The role of the media

The cases of Charlie Gard and Alfie Evans can be approached from many different viewpoints. In this chapter, the role of the media, including social media, and in particular the language used therein, will be the primary foci. The aim is to highlight the need for a more ethical press reporting, to reveal the need for greater transparency and accessible language regarding doctors' decisions, but also the opportunity for removal of discussion from the media gaze. Once in the media, the differing viewpoints became more intractable, with each side unable to persuade the other to a satisfactory, or at least mutually agreeable, solution. Mediation is therefore strongly suggested to avoid "framing and flaming" the debate. Mediation has been widely accepted in many spheres and offers at least the voicing of alternative interpretations in a neutral and confidential environment, and even of transforming the conflict – finding ways to redefine the issue.[1]

The methods used below to analyse the press include Framing theory as per Goffman[2] and linguistic analysis of mainly newspaper discourse, including quotes selected from relevant interested parties.[3] De-

---

[1] S. Littlejohn, "The Transcendent Communication Project: Searching for a Praxis of Dialogue," *Conflict Resolution Quarterly* 21 (2004): 337–359. doi: 10.1002/crq.66.

[2] E. Goffman, *Frame Analysis: An Essay on the Organisation of Experience* (Boston: Northeastern University Press, 1974).

[3] G. Tuchman, "Objectivity as Strategic Ritual: An Examination of Newsmen's Notions of Objectivity," *American Journal of Sociology* 77:4 (1972): 660-679. Gaye

spite the increase in various types of social media, newspapers continue to provide a major source of information for large sectors of the population, whether in hard copy or online versions. According to the National Readership Survey figures, *The Sun* and *The Daily Mail* are the most read UK wide newspapers.[4] It has been argued, and widely accepted, that reading a newspaper is a form of social ritual, and that social class and age influences the type of paper read.[5] Social class can be defined not by occupation, but from a person's way of being and living, their tastes, manners, arguably their propensity to emotion, their presentation of "self,"[6] including their vocabulary, and hence, even their perception and description of events. These are influenced by one's family, peers, and contacts, but also by one's reading. Hence, significant factors discussed below are the "reading" people undertake, including how the different newspapers and media posts appeal to readers, and how they "frame" and report issues.

However, it is also important to note that the use of social media, in the form of Twitter and Facebook, has increased exponentially in recent years. Research has shown how online communities can foster a shared sense of belonging and understanding, with shared cultural norms, arguably more powerfully than the press.[7] Twitter can similarly create a sense of shared community,[8] although it is more common to

---

Tuchman, *Making News: A Study in the Construction of Reality* (New York: The Free Press, 1978).

4 Ofcom, *News Consumption in the UK 2016*, p. 26. https://www.ofcom.org.uk/__data/assets/pdf_file/0016/103570/news-consumption-uk-2016.pdf (accessed June 20, 2018).

5 A. S. R. Manstead, "The Psychology of Social Class: How Socioeconomic Status Impacts Thought, Feelings, and Behaviour," *British Journal of Social Psychology,* (2018), https://doi.org/10.1111/bjso.12251 (accessed June 12, 2018).

6 E. Goffman, *The Presentation of Self in Everyday Life* (Garden City, NY: Doubleday/Anchor Books, 1959), 1-2.

7 N. K. Baym, *Tune in, Log On: Soaps, Fandom, and Online Community* (Thousand Oaks, CA: Sage, 2000). H. Rheingold, *The Virtual Community: Homesteading on the Electronic Frontier* (Cambridge, MA: The MIT Press, 1993).

8 Benedict Anderson had identified the importance of "imagined communities" and the role of the media in his influential work on nationalism. B. Anderson, *Imagined Communities: Reflections on the Origin and Spread of Nationalism* (London: Verso, 2006).

follow accounts of those whom one does not know personally.[9] The lack of personal knowledge does not, however, necessarily preclude a sense of shared community. Indeed, the reach of social media was such that Donald Trump, the US president at the time, also expressed his support for the parents in both the Charlie Gard and the Alfie Evans cases.[10]

This chapter argues that the involvement of the media is an unprecedented development in medical cases such as Alfie Evans and Charlie Gard, creating further complications for the resolution of such disputes. The analysis below draws heavily from actual press reports and illustrates how, as Goffman notes: "People, otherwise quite similar in their beliefs may yet differ in regard to a few assumptions, such as the existence of second sight, divine intervention and the like."[11] Indeed, he further argued that: "[b]elief in God … seems to constitute currently one of the largest bases of dissensus in our society concerning ultimate forces."[12]

For many parents of critically ill children, and noticeable in the cases of Charlie Gard and Alfie Evans and their supporters, their belief in divine intervention is often espoused, whether through pre-existing religious affiliation, hope, or desperation. This belief was evident in their posts and media discussion, fuelled by the involvement of the Pope in the debates surrounding both cases.[13] In the case of Charlie Gard, the Pope's intervention revealed "a change from the Holy See's position earlier in the week, when the Vatican's bioethics advisory panel had noted the need to do whatever helps a patient but also to some-

---

[9] A. Marwick and D. Boyd, "I Tweet Honestly, I Tweet Passionately: Twitter Users, Context Collapse, and the Imagined Audience," *New Media and Society* 13 (2011): 96–113.

[10] *The Guardian*, "Charlie Gard: Pope Shows Solidarity with Parents of Critically Ill 10-month-old" (July 2, 2017). https://www.theguardian.com/world/2017/jul/02/pope-shows-solidarity-with-charlie-gards-parents (accessed Jul 12, 2018).

[11] Goffman, *Frame Analysis*, 27.

[12] Ibidem, 28.

[13] *The Guardian*. "Charlie Gard."

times accept the limits of medicine."[14] The Pope similarly expressed support for Alfie Evans, against medical expertise.[15]

An Italian mother of a child with an allegedly similar condition to Charlie stated: "miracles happen, close to everyone, for God's grace for us is immense, the problem is we don't notice them. Our strength is in our Lord Jesus. 'We can do all this through Him who gives us strength,' the sentence is not ours, St. Paul said this, and he was right."[16] Not all involvement was free from ulterior motives, or discriminatory affiliations. Alfie's parents, for example, were supported by the Christian Legal Centre (CLC), a lobby group opposing abortion and gay rights.[17]

As noted above, with reference to Goffman's work, there is no doubt that people's beliefs often defy logic, appear impermeable to "reason" or to alternative narratives, and may potentially be harmful. Indeed, personal interests could lead to the dismissal of unpleasant facets relating to supporters, and to a rejection of a liberal democratic approach to alternative viewpoints for others. In the cases above, opposing viewpoints were often voiced vociferously and somewhat offensively, across the political or social spectrum. The British journalist, author, and public commentator Melanie Phillips, for example, strongly dismissed the supporters of Charlie Gard's parents:

---

[14] Ibidem.

[15] Research has indicated how cultural background, and context, affect one's interpretation of divine intervention and the existence of "miracles." Interpretation has tended to be unsympathetic to those expressing positive beliefs, seeing these as merely "delusional," without attempting to understand the context. See: I. Stahl and B. L. Jackson, "Belief in Miracles: Divine Intervention or Psychological Delusion?" *LUMEN-TCA* 21-22 (November 2014), Targoviste, Romania.

[16] G. Baggi, "Don't Give Up... Miracles Do Happen Mother of 'Italian Charlie Gard' says Son, 9, Given up for Dead as a Baby has a Happy Life of Painting and Chess" (5 July 2017). https://www.dailymail.co.uk/news/article-4668068/Italian-Charlie-Gard-s-mum-says-Miracles-happen.html (accessed July 13, 2018).

[17] *The Economist*, "The Pope's Role in the Plight of Alfie Evans was Regrettable" (29 April 2018).
https://www.economist.com/erasmus/2018/04/29/the-popes-role-in-the-plight-of-alfie-evans-was-regrettable (accessed July 14, 2018).

> I have never witnessed such concentrated ignorance, arrogance, stupidity, and unthinking cruelty as has been displayed by the American political right over the tragic case of Charlie Gard, … Here's the really wicked thing about all this . . . This campaign led the parents to believe that such pressure could change the court's mind. And so the parents were reinforced in their refusal to face reality.[18]

Crucial to this chapter's argument for the acknowledgment of alternative viewpoints in a "safe space," it is argued that refusing to acknowledge the existence and the power of another's beliefs, however different from one's own, was significant in the media portrayal of the dispute, and in the entrenching of viewpoints. Many different interpretations can be made regarding the initial use of the media: to change viewpoints; to garner political pressure; to merely achieve recognition of one's views. However, the effect was that once in the public arena, support for the parents was successfully achieved since they proved more astute and more "media savvy" in this respect. It quickly became clear that no media discussion could alter the other's viewpoint, and that in fact the involvement of the media proliferated and exacerbated the conflict. Issues were portrayed in social media in terms of "good" v "evil", truth v lies, as hospitals such as Great Ormond Street (GOSH) and the medical profession in general were perceived as related to both the latter. Posters, for example, proclaimed "GOSH, Tell the truth, where is the proof, release Charlie."[19]

The medical profession experienced and perceived aggressive pressure to allow the parents' wishes in both cases, in conflict with medical evidence that a different practice or treatment would be futile.[20] Doubts were widely espoused via media regarding the truth of medical evi-

[18] M. Phillips, *A Cruel and Ignorant Campaign* (July 26, 2017). https://www.melaniephillips.com/cruel-ignorant-campaign (accessed July 20, 2018).

[19] W. Gore, "Now Charlie Gard's Parents Have Ended Their Court Case, We Should Ask Why His Case Became The Media Circus It Did" (July 24, 2017). https://www.independent.co.uk/voices/charlie-gard-court-case-ruling-became-media-circus-a7857841.html (accessed March 1, 2019).

[20] Ibidem. See also N. Triggle, "Alfie Evans: Is 'Guerrilla Warfare' A Sign of Things to Come?" (April 30, 2018). https://www.bbc.co.uk/news/health-43898972 (accessed August 2, 2018).

dence.[21] Indeed, the more "evidence" supplied by the medical profession (within the limits of professional ethics and confidentiality), the more the parents claimed they, and their children, were victims of an uncaring system, that their own wishes were being undermined, and the children were effectively imprisoned.[22] It has long been accepted that perceptions of "the truth" depend not upon evidence, but upon interpretation, and in particular, confirmation bias, influenced by context and one's own background.[23] Each "side" inevitably felt their position was confirmed by the "evidence" they garnered. The effect was devastating, as the paediatrician Dr. Jayaram expressed:

> It is sad to hear about cases that end up having to go to court; I wonder how the gulf of understanding between the families and the teams looking after their children has been allowed to open up. A judge's decision on what should happen is an unsatisfactory outcome for everyone compared to a plan negotiated and designed together.[24]

II. Framing of media stories

Those involved selectively chose evidence and saw this as justification for their views – examples of confirmation bias and can be related to the process of "framing." Framing, as defined by Goffman, refers to the frameworks or schemata employed to interpret events. Since these can be subtle and rely on concepts of perceived "common sense," their influence is often overlooked.[25] However, what is "accepted" as "common sense" is itself a normative process, that can differ between groups, but with a dominant perception framing any issue. Hence, the very notion of, and appeal to, "common sense" carries with it various

[21] W. Gore, "Now Charlie Gard's Parents."

[22] N. Triggle, *Alfie Evans: Is 'Guerrilla Warfare'.*

[23] R. S. Nickerson, "Confirmation Bias: A Ubiquitous Phenomenon in Many Guises," *Review of General Psychology* 2:2 (1998): 175-220.

[24] R. Jayaram, "As A Paediatrician, I Can Tell You That The Case Of Charlie Gard Isn't As Black and White As It Seems," *The Independent* (July 6, 2017). https://www.independent.co.uk/Voices/charlie-gard-gosh-great-ormond-street-hospital-life-support-pope-donald-trump-a7827276.html (accessed June 10, 2018).

[25] Goffman, *Frame Analysis.*

interpretations, not least one relating to potentially conflicting public or expert opinions. For the Romans, "common sense" was indeed that of the "common" people, and therefore seen as "vulgar." "Vulgar law" was understood as not official or classical Roman law, but the law practised by the peoples and their conception of what was required.[26] This is clearly not the Aristotelian view of common sense (*koine aisthesis* or *sensus communis*) as a higher-order perception that humans uniquely possess.[27] Nor is it the "common sense" defined by the Marxist Gramsci as reflecting and imposing the interests of dominant classes.[28] In the 21st century, we are theoretically attuned to the infinite variety of viewpoints that may exist, and yet opposing views tend to be framed within these three unequal concepts, even if subliminally. The legal and medical systems portray themselves as embodiments of rationality and reason,[29] and hence, a specific interpretation of "common sense." Encompassing differing opinions has proven problematic, but normally concerns refusal of treatment for example, for religious reasons, with some authors choosing to define these as "non-rational."[30]

Every day, these concepts of "rationality" and accompanying "legal and medical paternalism" are usually accepted. Parents' demands for treatment have been refused on the grounds of medical futility,[31] medi-

[26] H. Brunner, *Zur Rechtsgeschichte der Römischen und Germanischen Urkunde* (Berlin, 1880), 113; see also 139; and idem, *Deutsche Rechtsgeschichte* I (Leipzig, 1887), 255, 2nd ed. 1906, 377, cited in Boudewijn Sirks (Hrsg.). *Aspects of Law in Late Antiquity: Dedicated to A.M. Honoré on the Occasion of the Sixtieth Year of his Teaching in Oxford.* (Oxford: Souls College, 2008): S. 35-53.

[27] *Aristotle's Psychology: A Treatise on the Principle of Life* (De Anima and Parva Naturalia), ed. W. Hammond, trans. William Hammond (Cambridge, MA, Swann Sonnenschein, 1907), 204. Digitised for Google Books on June 6, 2006.

[28] *Gramsci and Marxist Theory*, ed. C. Mouffe (London: Routledge, 1979).

[29] R. Anthony-Pillai, "Alfie Evans Case: Are the Notions of Best Interests and Futility Just Paternalism Rebranded?" *British Medical Journal* 361(2018) (Published June 1, 2018) doi: https://doi.org/10.1136/bmj.k2436.

[30] M. Stauch, "Rationality and the Refusal of Medical Treatment: A Critique of the Recent Approach of the English Courts," *Journal of Medical Ethics* 21 (1995): 162-5.

[31] L. J. Schneiderman, N. S. Jecker, A. R. Jonsen, "Medical Futility: Its Meaning and Ethical Implications," *Annals of Internal Medicine* 112 (1990): 949-54.

cal standards,[32] patient suffering, lack of net benefit to the patient,[33] or cost.[34] Numerous life and death cases are decided by various courts, including the European Court of Human Rights, without the accompanying public animosity revealed by the two cases of Charlie Gard and Alfie Evans. Indeed, public opinion is rarely invoked to assess legal or medical rationality. Doctors and charities stress that cases like Charlie Gard and Alfie Evans are exceptions.[35] However, doctors state that there are "at least 49,000 children and young people in the UK with illnesses that may cut their lives short."[36] Therefore, the potential for future disputes exists, especially if the media become involved. The cases of Charlie Gard and Alfie Evans were highly publicised, across multiple media, and involved very young children, and young parents. It has been argued that medical futility cases result in mistrust between doctors and patients or their families[37] and that these two cases, in particular, may set a worrying precedent in that medical and legal decisions will be tried by the media, resulting not in solution, but disappointment and increased mistrust.[38]

The difficulty for the medical profession in such cases was highlighted by the paediatrician Dr. Jayaram:

---

[32] J. J. Paris, M. D. Schreiber, M. Statter, et al, "Beyond Autonomy – Physicians' Refusal to Use Life-Prolonging Extracorporeal Membrane Oxygenation," *New England Journal of Medicine* 329 (199): 354-7.
[33] S.H. Miles, "Informed Demand for 'Non-Beneficial' Medical Treatment," *New England Journal of Medicine* 325 (1991): 512-5.
[34] F. D. Moore, "The Desperate Case: CARE (Costs, Applicability, Research, Ethics," *Journal of the American Medical Association* 261 (1989): 1483-4.
[35] C. Burns, "Charlie Gard's Parents Want 'Charlie's Law'," Health Correspondent, BBC News (June 20, 2018).
https://www.bbc.co.uk/news/health-44334306 (accessed August 2, 2018).
[36] Ibidem.
[37] Lawrence J. Schneiderman, Nancy S. Jecker, and Albert R. Jonsen, "Medical Futility: Response to Critiques," *Annals of Internal Medicine* 125 (1996): 669-674.
[38] J. Sharman, "Stop Attacking Doctors and Read the Full Charlie Gard Court Judgments, Paediatrician Tells Social Media Users," *The Independent* (July 8, 2017). https://www.independent.co.uk/news/uk/home-news/charlie-gard-mother-connie-yates-donald-trump-pope-offer-help-not-in-pain-terminally-ill-baby-boy-a7828601.html (accessed September 2, 2018); R. Heywood. "Parents and Medical Professionals: Conflict, Cooperation, and Best Interests," *Medical Law Review* 21:1 (2012), 29.

> As human beings, we like things to be black and white, good guys versus bad guys, the establishment versus the little man. Unfortunately, medicine inevitably operates in shades of grey and there is rarely a 100 percent "right" thing to do. Every decision is a balance of potential benefit against potential harm. As a result of the binary media coverage, many people have developed fixed opinions on the situation on one side or the other, which they express passionately and loudly.[39]

The media, and particularly the tabloid press or social media, are not traditionally known as fora for rationality, but often rely upon sensationalism, emotion, and simplification: the latter is seen as a lesser form of news journalism in much academic writing.[40] However, recent writing has questioned this value-driven interpretation, arguing that tabloid journalism represents the public interest.[41] As early as 1997, Franklin argued that all media was becoming more like tabloid media.[42] Thompson further argued that the mediated public sphere is one where struggles for visibility now occur.[43] The two legal cases have proven highly emotional, and widely publicised, with most criticism focussing upon the effect on the child and his parents. Many critics of the way the cases were handled supported and emphasised the "rights of parents," in particular, to decide what medical treatment their children should receive. A common theme across social media, widely stated, and similarly believed by many, was that the parents were the ones fighting for what was best for their child.[44]

---

[39] R. Jayaram, "As a Paediatrician."

[40] *Tabloid Tales. Global Debates Over Media Standard*, ed. C. Sparks and J. Tulloch (Kindle Edition: Rowman & Littlefield, 2000).

[41] H. Örnebring and A. M. Jönsson, "Tabloid Journalism and the Public Sphere: A Historical Perspective on Tabloid Journalism" (published online: 17 Feb. 2007), 283-295. https://www.tandfonline.com/doi/full/10.1080/1461670042000246052?src=recsys (accessed November 20, 2018).

[42] B. Franklin, *Newszak & News Media* (London: Arnold, 1997).

[43] J. B Thompson, *The Media and Modernity: A Social Theory of the Media* (Oxford: Polity Press, 1995), 247.

[44] *The Guardian,* "Connie Yates: We Will Always Know We Did The Very Best For Our Son Charlie Gard" (July 24, 2017).

Alternatively, doctors treating Charlie Gard and Alfie Evans relied upon their medical expertise, and argued that the children's conditions were irreparable, further intervention to be futile, and not in the children's best interests. However, beyond the issue of "rights" and "interests," the debates were framed in two very different types of language, crucially affecting the interpretation of the actual situations. Such interpretations have serious implications for the public perception of medical professionalism, beyond these two particular cases.

Both cases shared *emotive* information and language through social media, coming from the parents, not from the medical viewpoints. For instance, Charlie Gard's parents characterised the hospital and its lawyers as "evil," before rushing out of the courtroom when medical evidence was given.[45] Medical experts cited the facts of Charlie's condition, that he had a rare form of mitochondrial disease, and may have been suffering.[46]

Undoubtedly this media information did inform and capture people's attention and mobilised many people in support of the parents, resulting in a form of activism. Research upon Twitter references to Great Ormond Street Hospital revealed an overwhelming use of the words "murder" and "killed."[47] As Das argues, the analysis of words used to suggest devastation and loss revealed the top ten to be: devastated, heart-breaking, cruel, death panel, loss, tragic, and threaten.[48]

This chapter argues that the use of social media for such purposes requires some ethical dimension, considering the consequences of selective "information sharing" and, moreover, what should be shared, and how. Medical staff suffered abuse at both Great Ormond Street Hospital and Alder Hey Hospital, with Merseyside police investigating

https://www.theguardian.com/uk-news/2017/jul/24/connie-yates-we-will-always-know-we-did-the-very-best-for-charlie-gard (accessed July 28, 2017).

[45] ITV Report, "Dad Yells 'Evil' as Lawyer Reveals Charlie Gard's Latest Scan Makes 'Sad Reading'." https://www.itv.com/news/2017-07-21/dad-yells-evil-as-lawyer-reveals-charlie-gards-latest-scan-makes-sad-reading/ (July 21,2017) (accessed August 3, 2018).

[46] Ibidem.

[47] R. Das, "The Charlie Gard Twitterstorm: A Violent and Negative Impact (New Research)," https://blogs.lse.ac.uk/polis/2017/08/04/the-charlie-gard-twitterstorm-a-violent-and-negative-impact-new-research/.

[48] Ibidem.

allegations that patients and staff at the latter had been intimidated.[49] Arguably, such behaviour was not solely or even largely due to the information shared, but more *the manner in which the situation was manipulated by parents and the media, particularly the tabloid press.*

Studies on the language of the press have previously shown significant differences in use in the tabloid and broadsheet press. The former contains more emotive and often inflammatory language, for example "killing a child," and reducing the case to a "soap opera."[50] Social media, for example upon the internet, and Twitter, similarly rely upon soundbites and appeals to the emotional. Indeed, the 2002 Nobel Prize winner, psychologist Daniel Kahneman, in his book *Thinking, Fast and Slow*, demonstrated how people use two very different systems in their minds: Intuitive and Automatic, and Lazy and Calculative.[51] The former utilises gut instinct, emotions, and unconscious decision-making: the latter slow, conscious, and rational thinking. Anger is one of the strongest emotions, and hence, it can be activated quite easily in media campaigns, such as the two cases herein. Whilst the rational can override the emotional, in reality, people tend to rely more upon the first system. Social media draws upon this psychological reality to its own advantage, whereas the medical profession, arguably, tends to be centred in the second system, and thus more limited in its social appeal potential.

The academic studies above on "framing" and "thinking" processes have not traditionally considered medical cases, although some recent works have done so.[52] They have, however, focussed upon the social

---

[49] BBC News, "Alfie Evans: Legal Battle Toddler Dies" (April 28, 2018). https://www.bbc.co.uk/news/uk-43933056 (accessed August 4, 2018).

[50] C. Mortimer, "Charlie Gard Medic Hits Out at 'Soap Opera' Surrounding the Court Case and Meddling of Politicians," *The Independent* (August 5, 2017). https://www.independent.co.uk/news/uk/home-news/charlie-gard-court-case-great-ormond-street-hospital-high-court-battle-connie-yates-chris-gard-a7878026.html (accessed August 9, 2018).

[51] D. Kahneman, *Thinking, Fast And Slow* (London: Penguin Books Ltd., 2012).

[52] T. V. Perneger and T. Agoritsas, "Doctors and Patients' Susceptibility to Framing Bias: A Randomized Trial," *Journal of General Internal Medicine* 26:12 (2011): 1411–17.
doi:10.1007/s11606-011-1810-x. PMC 3235613. PMID 21792695.

construction of social phenomena, and social interaction.[53] Thus, this chapter argues that not only what information is given, but the language used, and the context, have a significant effect upon perceptions of "reality" and have a significant impact, proving crucial in these medical cases. Regarding what information was available via the press, it was stated that:

> ... journalists and the SMC (Science Media Centre) found it very difficult to find medics, paediatricians and palliative nurses who could speak. Once again, the voice of the medical community was represented by very few individuals. And so once again the level of information available to the public was woefully imbalanced... Managers and executives in hospitals, NHS Trusts, and medical institutions were preventing their staff from speaking to the media about the Alfie Evans story. Time and again I was told "I've been told I can't give you a quote," "I am being muzzled," "I agree experts need to be heard but my manager has put us all on lockdown." These comments did not come from hospital juniors, but from professors, senior medics and leading paediatric nurses.[54]

No doubt, medical professionals and managers feared reprisals and were risk averse regarding engaging with the press. However, the effect was that the subsequent press coverage was severely imbalanced, and in the parents' favour.

Moreover, even when messages were given by the medical profession, these appeared tied to mainly managerial and medical rhetoric, defining the condition in medical terms, "mitochondrial disease" and "best interests." In contrast, the legal profession, still constrained by protocol, was able to state, as per Mr Justice Francis, that the decision to allow the cessation of life support was made: "with the 'heaviest of hearts' but with 'complete conviction' (that) it is in Charlie's best inter-

---

[53] Goffman, "Frame Analysis"; *Media Effects: Advances in Theory and Research*, ed. J. Bryant & M. B. Oliver (New York: Routledge, 2008); T.V. Perneger, T. Agoritsas, "Doctors and Patients."

[54] T. Sheldon, "There will be Another Alife Evans: Experts Need to Engage" (May 10, 2018). http://www.sciencemediacentre.org/there-will-be-another-alfie-evans-experts-need-to-engage/ (accessed August 4, 2018).

ests."[55] In contrast, the parents and the mainly tabloid media were able to utilise often highly emotive and inflammatory images. As Wittgenstein argued, "the meaning of a word is its use in the language."[56] Thus, meaning is not a concrete reference, "language (is) a game: the game of language." [57]

Social media proved more adept at playing the game and manipulating public opinion via language and images. Both sets of parents used selective information and photographs of their children not necessarily related to the child's current situation. Charlie Gard's parents, for example, used photographs of a two-week old Charlie, thus taken before his condition was diagnosed at eight weeks, when he was admitted to hospital with encephalomyopathic mitochondrial DNA depletion syndrome. The medical name would mean little to the layperson, particularly since this was a rare disease caused by mutations in gene RRM2B inherited from both parents. Indeed, it was widely reported, particularly in the tabloid press, that only 16 people in the world have been diagnosed with this condition.[58]

The images supplied via the media gave little or no indication of the actual serious condition of Charlie. Instead, these images often showed earlier images or those suggesting merely sleep, accompanied by children's comforters, such as teddy bears. The language of the media thus includes visual imagery. Indeed, a teddy bear became an image associated with Charlie Gard, worn by his parents, and attached to the gates of the Royal Courts of Justice.[59]

---

[55] Mirror News, "His Legacy Will Never Die': Charlie Gard Passes Away after Capturing Nation's Hearts in 11 Previous Months" (July 28, 2017). https://www.mirror.co.uk/news/uk-news/charlie-gard-baby-boy-dies-10891696 (accessed August 4, 2018).

[56] L. Wittgenstein, *Philosophical Investigations* translated by G.E.M. Anscombe (Oxford: Basil Blackwell, 1963): Sect. 43.

[57] Ibidem.

[58] Reported in Alison Smith-Squire and Sophie Roberts, "If We Were British Our Son Would Be Dead" (*The Sun*, April 13, 2017). https://www.thesun.co.uk/living/3318065/arthur-olga-estopinan-arturito-charlie-gard-court-case/ (accessed November 20, 2018).

[59] The image can be seen within this online article: C. Domonoske, "Terminally Il Baby Charlie Gard Will Move to Hospice Care to Die," https://www.kut.org/post/terminally-ill-baby-charlie-gard-will-move-hospice-care-die (accessed November 20, 2018).

Furthermore, both sets of parents denied the medical definitions of their children's conditions, and both vociferously argued that potential treatment abroad could be "life-saving."[60] Medical professionals stated that both Charlie and Alfie had severe brain damage, that Charlie could not move his limbs or breathe unaided, and that no cure existed.[61]

The parents used many types of descriptions for the evoking viewpoints to which they disagreed, partly depending upon the narrator. When brain scans were used in court to reveal the extent of Charlie's situation, his parents responded by calling the lawyer "evil" for revealing this.[62] Alfie Evans[63] was diagnosed by scans at Alder Hey Hospital as having "catastrophic degradation of his brain tissue."[64] The hospital argued that further treatment for Alfie was not only futile but also "unkind and inhumane,"[65] Alfie's parents responded by "heavily (criticising) medical staff, with Mr Evans suggesting his son was a 'prisoner' at the hospital and had been misdiagnosed."[66] Michael Mylonas QC, representing the hospital, said: "One of the problems of this case is they [Alfie's parents] look at him and, barring the paraphernalia of breathing and feeding, he's a sweet, lovely, normal-looking boy who opens his eyes, [and] will smile..."[67] Whereas the hospital asserted that any movements by the child were "spontaneous seizures as a result of touching."[68]

---

[60] N. Debnath, "Charlie Gard: Mother Connie Yates Makes Heartbreaking Plea to Try New Treatment" (July 7, 2017). https://www.express.co.uk/showbiz/tv-radio/825809/Charlie-Gard-mum-Connie-Yates-Chris-Gard-Good-Morning-Britain-ITV (accessed August 15, 2018).

[61] BBC News, "Who was Alfie Evans and What Was the Row over His Treatment?" (April 28, 2018). https://www.bbc.co.uk/news/uk-england-merseyside-43754949 (accessed July 11, 2018).

[62] G. Heffer, "Timeline: Parents' Battle to Save Charlie Gard," (August 4, 2016). https://news.sky.com/story/timeline-parents-battle-to-save-charlie-gard-10914755 (accessed July 6, 2018).

[63] BBC News, "Alfie Evans: Legal Battle Toddler Dies" (April 28, 2018) https://www.bbc.co.uk/news/uk-43933056 (accessed July 10, 2018).

[64] Ibidem.

[65] Ibidem.

[66] Ibidem.

[67] BBC News, "Who Was Alfie Evans."

[68] Ibidem.

Examples of the *constraints of managerialism* within the NHS can be seen in the comments of medical ethicist Dr. Daniel Sokol. Indeed, Dr. Sokol's comment neatly summarises the language outside of the medical profession, but also the limitations placed upon an NHS response. He stated that the approach of the parents:

> smacks of "guerrilla warfare" where threats, insults, intimidation, and distortion of facts were used to put "unbearable pressure" on the health service… In both cases, parents and their supporters were able to use social media to make their points. And when faced with emotional pleas from parents - … it is almost impossible for the NHS to provide the counter-argument while ensuring it maintains patient confidentiality, remains respectful to families and protects its staff. It creates an environment where the NHS can be painted as uncaring and the weight of expert opinion can be ignored by the public.[69]

Clearly, however, Dr. Sokol was not averse to evoking the language of warfare when discussing the parents and the media coverage. Undeniably the literal, metaphorical, and visual imagery of warfare was overt in social media. The metaphorical imagery is common in discourse regarding, particularly, cancer (with both negative and positive implications).[70] However, in the cases above, the metaphors involved more than one connotation, suggesting both a fight against disease in general, but also against the medical profession, and the hospitals involved. The public were visibly and vociferously engaged in the parents' legal fights with the hospitals. Charlie Gard's parents used "Charliesfight" and "charliesarmy" in their banners.[71] Alfie Evans' father referred to: "My gladiator laid down his shield" on social media.[72] Web pages and Twitter feeds used the titles: Charliesfight.org and #charliesarmy. Charlie was proclaimed to be "still fighting," despite there being little indication

---

[69] N. Triggle, "Alfie Evans: Is 'Guerrilla Warfare.'"

[70] D. J. Hauser and Norbert Schwarz, "The War on Prevention Bellicose Cancer Metaphors Hurt (Some) Prevention Intentions," *Journal of Experimental Social Psychology* 48 (2014): 948-952.

[71] Image available here: BBC News, "Alfie Evans Legal Battle."

[72] Ibidem.

or possibility of him voluntarily exerting such action.[73] The metaphors of army and fighting thus reaffirmed not only the image of conflict and war, against a disease, or medical condition but also of incompatibility and intractability regarding the medical advice. Arguably, such negative connotations severely hindered the public understanding and ultimate discussion and resolution of these difficult cases.

Ms. Yates, Charlie's mother, stated on national television and other media that she had spoken to doctors all over the world, with five offering treatment[74] and had endlessly researched a possible cure for Charlie's condition. Indeed, Charlie Gard's case achieved not only worldwide notoriety but the support of "celebrities" such as Sam Faiers, Billi Mucklow (reality television stars), television presenter Gaby Roslin, and actress Linda Robson. Interest extended beyond television with the UK Prime Minister Theresa May, and, as noted above, the Pope and Donald Trump also supporting Charlie Gard's parents. None of these supporters were medically or legally qualified. Moreover, the media itself subjectively manipulated opinion. Hurley argues, for example, that the *Daily Mail* juxtaposed images from the Charlie Gard case next to a headline relating to a different recent event – "Manslaughter" and the Grenfell tragedy.[75]

Whilst Great Ormond Street Hospital did supply statements regarding the ongoing court decisions, the language was subdued and respectful, albeit given by "a spokesperson," and hence not someone with whom the public could identify. On the final decision by the High Court to not pursue further treatment, the hospital gave a public statement. The message paid tribute to the family involved, yet also reiterated the hospital's role for countless others. Arguably this understandable

---

[73] Ibidem.

[74] N. Debnath, "Charlie Gard."

[75] R. Hurley, "How a Fight for Charlie Gard Became a Fight against the State," *British Medical Journal* (2017): 358, j3675*BMJ* (1 August 2017). The "Grenfell Tragedy" occurred on 14 June 2017. This refers to a fire at the 24-storey Grenfell Towerblock of flats in North Kensington, West London. It caused 72 deaths, including those of two victims who later died in hospital. More than 70 others were injured, and 223 people escaped. It was the worst UK residential fire since the Second World War. Its causes are under investigation, with some in the press believing that gross negligence manslaughter could be an appropriate consideration.

emphasis may have suggested to some that the reputation of the hospital was a primary concern. In the light of the negative publicity surrounding these cases this is perhaps only to be expected:

> A spokesperson for Great Ormond Street Hospital said: We have always paid tribute to the extraordinary courage and strength of Connie Yates and Chris Gard and their devotion to their son. Today our hearts go out to them as they face the most difficult decision that any parent should ever have to make.
>
> Our staff share the parents' sadness over Charlie, whom they have cared for with dedication during the months he has spent at Great Ormond Street Hospital. We were heartened to hear their appreciation of our staff's care in their recent public statement.
>
> We seek to care for and support every parent who brings their child to Great Ormond Street Hospital and it is a source of great regret in the very rare cases when an agreement cannot be reached about what is in the child's best interests.
>
> This has been an incredibly complex case, going to the heart of medical ethics and when, and if, experimental treatment is acceptable for children suffering from rare diseases. We would like to acknowledge all those who have contributed their expertise and advice.[76]

III. Conclusion: the need for mediation

Above has discussed the role of the media in medical futility cases. It has highlighted the apparent "intractability" that can occur once such cases enter the media gaze. Below concludes by discussing the potential for mediation to avoid "framing" and "flaming" the debates in such difficult cases, and for ideally enabling a mutually acceptable conclusion. At the very least, mediation gives the opportunity for all sides to state their views, in a confidential environment. To avoid, or ameliorate, similar situations to the Charlie Gard and Alfie Evans cases in the

---

76 "Latest Statement on GOSH Patient Charlie Gard" (July 24, 2017). https://www.gosh.nhs.uk/news/latest-press-releases/gosh-position-statement-issued-high-court-24-july-2017 (accessed July 25, 2017).

future, this chapter argues for the need to recognise that there may be strongly held, but potentially incompatible, beliefs by the parents and the medical profession regarding the treatment of young children. Mediation is suggested as a method to promote if not a completely satisfactory result for all parties, one that is capable of achieving a remotely acceptable result, without the risk of mass media involvement and public antagonism.

Ideally, the format would be a neutral environment, utilising an impartial mediator. The current dependency upon the law to resolve disputes that have become intractable is far from ideal. Since clinicians are more constrained in the information they can reveal to the press, and the methods of doing so, the confidential and impartial forum that mediation offers would be preferable. Much of the antagonism directed towards the hospitals in the cases above appears to emanate from a sense of not being listened to, and an inability to express alternative viewpoints.[77] The benefits of mediation are often espoused as providing a forum and an atmosphere conducive to mutual understanding and cooperation. The ultimate aim is a mutually-agreed resolution, determined by the parties. Mediation is stated to be a cost-efficient method, not merely in financial terms.[78]

As Heywood states, the costs associated with a breakdown in the relationship between clinicians and parents can be financial, emotional, and physical. Clinicians are, in a very real sense, mediators. Treatment is an enterprise that is dependent upon cooperation and this necessitates a relational approach. Whilst the duty to communicate with and consult parents is clear in both law[79] and ethics,[80] the waters are muddied by inconsistent ethical guidance on unproven, innovative treatment and latterly by inconsistencies in the legal approach.[81] The involvement of the ECtHR in *Glass v UK*, for example, focussed upon legal avenues,

---

[77] C. Burns, "Charlie Gard's Parents Want 'Charlie's Law'."

[78] H. Morreim, "Conflict Resolution in the Clinical Setting: A Story Beyond Bioethics Mediation," *Journal of Law, Medicine and Ethics* (Research Article First Published December 1, 2015). https://doi.org/10.1111/jlme.12324.

[79] *Glass v United Kingdom* (App no 61827/00) [2004] 1 FCR 553.

[80] R. Heywood, "Parents and Medical Professionals: Conflict, Cooperation, and Best Interests," *Medical Law Review* 20:1 (2012): 29-44.

[81] E. Cave, E. Nottingham, "Who Knows Best (Interests)? The Case of Charlie Gard," *Medical Law Review* 26:3 (2018): 500-513.

and the need for doctors to seek authorisation from the courts to overrule parents' objections, when both sides felt they were acting in the "best interests of the child." The mother in *Glass* sought an unsuccessful judicial review of the doctor's treatment and similarly unsuccessful criminal action. However, the ECtHR found a breach of Article 8 for failure to seek legal backing prior to action being taken. It has been argued that in the Charlie Gard case:

> Professional guidance from the GMC and Royal College of Paediatrics and Child Health raised doubts as to the ethics of sanctioning the unproven therapy and in law, it was questionable on existing legal precedent that it would satisfy either the best interests test or the *Bolam* standard of reasonable care. In light of this, mediation could not have focussed on finding a compromise, for there was no such middle path.[82]

Undoubtedly, sanctioning mediation as a method to consider the wishes of the parents, including the use of unproven therapy, provided the risks were understood, could be a departure from the emphasis upon ethical and legal guidelines, and requires a rethinking of what, and who decides, is in the child's best interests. However, this chapter argues that such an approach could be considered in the future. The final mediation decisions would be binding, but confidential, unless a public interest case existed, and must respect any human rights issues.

Without the possibility of a "public interest" clause, such confidentiality would mean that these cases would not set a precedent for the future which could be considered a negative outcome for developments in this area. Indeed, a number of critics (e.g. Fiss; Nader; and Edwards) have made such comments, accusing mediation of "sweeping injustices under the carpet by dealing with each individual dispute privately, so that bad practice attracts no negative publicity and no precedents are set for the future."[83]

---

[82] Ibidem.

[83] C. Irvine, "Mediation Confidentiality: Limitations and a Proposal" (2012). http://mediationblog.kluwerarbitration.com/2012/09/12/mediation-confidentiality-limitations-and-a-proposal/ (accessed August 5, 2018). O. M. Fiss, "Against Settlement," 93 *Yale Law Journal* 1073 (1984): 1085–87; L. Nader, "Disputing Without the Force of Law," 88 *Yale Law Journal* 998(1979):

Conversely, as this chapter argues, once in the public sphere, the wishes and perceptions of the parents appear paramount and are most often espoused, in highly emotional terms. In this arena, the medical profession appears cold and distant, constrained by confidentiality and medical jargon. In the forum of mediation, both parties would have the opportunity to voice their views, without the opportunity for the media to "frame and flame" the debate. As Mr Justice Francis in the Charlie Gard case stated: "It is my clear view that mediation should be attempted in all cases such as this one even if all that it does is achieve a greater understanding by the parties of each other's positions."[84]

Since the death of Charlie Gard, his parents have used the media to campaign for *Charlie's Law*: "One key part of the law would be having independent medical mediators involved in disagreements as early as possible."[85] In their media campaign, they quoted Great Ormond Street Hospital as agreeing that: "mediation is one of the important ways in which we work with families to reach agreements, in the best interests of the child."[86] However, Charlie's parents stated that they had asked for mediation in the months prior to Charlie's death, but that this had been refused.[87] As this chapter argues, one key issue in such cases is the need for both parties, and in particular, the parents, to feel that their voices are being heard. If this is not undertaken via mediation, it is highly likely that a media campaign will provide that space. As Charlie's father stated: "If mediation had taken place with an independent mediator, we would have felt our voices were being listened to."[88]

Whilst Great Ormond Street Hospital argued that it offered mediation in the final few days of Charlie's life,[89] clearly, this was too little,

1007–08). See also H. Edwards, "Alternative Dispute Resolution: Panacea or Anathema?" *Harvard Law Review* 99 (1986): 668-684, 671-72, 675-82.

[84] Mr Justice Francis, quoted in *The Guardian,* "Charlie Gard's Short Life Pitched Parents into Turmoil and Grief," (July 28, 2017). https://www.theguardian.com/uk-news/2017/jul/28/charlie-gards-short-life-pitched-parents-into-turmoil-and-grief. See, also, for example, https://www.medicalmediation.org.uk (accessed August 10, 2018)

[85] C. Burns, "Charlie Gard's Parents Want 'Charlie's Law'."

[86] Ibidem.

[87] Ibidem.

[88] Ibidem.

[89] Ibidem.

too late, to prevent the involvement of the media, with the subsequent effects discussed above. Lessons for the future should include the timing of the offer of mediation. The final word is given to Sarah Barclay, Director of the Medical Mediation Foundation, who mediates between five and ten parents-doctors conflicts a year: "Sometimes it feels like there's a bit of a battle developing." She explains: "Mediation is always surprising in the way it can break through really intractable conflicts."[90]

[90] Ibidem.

Chapter 5

Resource-intense Treatments in a Resource-finite Environment

Richard Wai Ming Law
University Hospitals Morecambe Bay NHS Foundation Trust, UK

I. Introduction

Resource allocation and decisions concerning the rationing of medical treatment are controversial issues for the National Health Service (NHS).[1] In the UK, the wholly taxpayer-funded NHS healthcare system provides healthcare to the general public free at the point of use. Since its founding in 1948, the service promulgated several primary operational objectives coordinated by the Department of Health.[2] These include nationwide public financing of healthcare services, control of healthcare distribution across the country, appropriate planning, and coordination of healthcare delivery.[3] The annual budget allocated to the NHS by the UK government for the year 2017-18 was £124.7 billion.[4] Given that the financial sustainability of the NHS is clearly important

---

[1] The National Health Service was established on 5th July 1948 following the Beveridge Report 1942. The report proposed "any plan to establish a national social security system would include a national health service for prevention and comprehensive treatment, which would be available to all members of the community."

[2] Department of Health, *The Government's Mandate to NHS England for 2017-18.* (London: Department of Health, 2018), https://assets.publishing.service.gov.uk/government/uploads/system/uploads/attachment_data/file/692140/NHSE_Mandate__2017-18_revised.pdf (accessed November 30, 2018).

[3] F. Mosley and S. Lockwood, "Prioritizing Clinical Care in the Cash-Strapped NHS," *Surgery* 36:9 (2018): 491-496.

[4] The Kings Fund, *The NHS Budget and How it has Changed* (London: The Kings Fund, 2018). https://www.kingsfund.org.uk/projects/nhs-in-a-nutshell/nhs-budget (accessed November 29, 2018).

to all patients engaging with the health service, resource allocation affects how it utilises its substantial budget to provide universal medical coverage.[5] How best these resources are allocated to achieve maximal utility for the health of the nation presents a multi-layered problem[6] - the imperative for the service to be fair, just, efficient, and cost effective.[7]

Nationally, politicians and executive healthcare managers may determine how resources are allocated to various healthcare institutions across the country.[8] At regional and local levels, Clinical Commissioning Groups (CCG), which are ultimately accountable to the Secretary of State for Health, decide available treatments for each region of the country. These decisions are based on regional demographic data, patient clinical outcomes, and previous spending.[9] These factors contribute to observable regional differences in healthcare throughout the country and variations in service delivery between different hospitals. At the local level, CCGs[10] were set up by the Health and Social Care

---

[5] In Lord Darzi's evaluative report – High Quality Care for All, "Clinicians must be partners in care delivery with individual and collective accountability for the performance of the health service and for the appropriate use of resources in the delivery of care." See: Department of Health, *High Quality Care for All: NHS Next Stage Review Final Report* (London: The Stationery Office, 2008), 60.

[6] P. Ellis, "Resource Allocation: Exploring Concepts of Justice and Utility," *Journal of Kidney Care* 3:1(2018): 42-3.

[7] L. Williams, S. Robinson, H. Dickinson, *Rationing in Health Care: The Theory and Practice of Priority Setting* (Bristol: Policy Press, 2012). See also M. Luchetti, "Intensive Care Resource Allocation: When Difficult Choices Have to be Made," *British Journal of Medical Practitioners* 6 (2013): 633.

[8] UK Parliament, *Funding and Commissioning Information* (London: UK Parliament, 2015).
https://publications.parliament.uk/pa/cm201415/cmselect/cmwelaf/404/40406.htm (accessed 7 November 2018).

[9] NHS Commissioning Board, *Commissioning Policy: Ethical Framework for Priority Setting and Resource Allocation April 2013 Reference NHSCB/CP/01* (London: NHS Commissioning Board, 2013). https://www.england.nhs.uk/wp-content/uploads/2013/04/cp-01.pdf (accessed November 8, 2018).

[10] There are currently 207 CCGs across England. The control of the NHS in Scotland, Wales, and Northern Ireland are devolved matters to each respective Governments.

Act 2012 and are clinically led statutory NHS bodies.[11] Each CCG is led by a governing body comprising of healthcare professionals and lay individuals. In England, CCGs control approximately 60% of the NHS budget and this is used to finance most hospital-based and primary healthcare.[12] Notwithstanding ever-increasing healthcare spending, novel medical technologies, treatments, and therapies will always create cost pressures.[13]

With an ageing population and increasing numbers of individuals with multiple comorbidities, patients frequently present to the health service with clinical problems that are often difficult to manage and expensive to treat. The majority of NHS expenditure is utilised for the care of adults in old age,[14] though the healthcare of children is an important aspect of a comprehensive and universal healthcare system.[15] The high profile cases of Charlie Gard and Alfie Evans illustrate that children *(as a patient cohort)* may present with their unique clinical problems that require complex medical care; and that expensive treatments are not necessarily merely reserved for elderly patients with chronic disease. Unfortunately, acute care hospitals are not designed to provide long-term care for hospital-dependent patients with complex care needs. Dedicating limited resources to short term treatments in patients who are fully dependent on hospitalised care without long term plan-

[11] Health and Social Care Act 2012 section 26. Note that in subsection 14Q, each CCG must "exercise its functions effectively, efficiently and economically."

[12] National Health Service Act 2006. Under section 14R (as amended by the Health and Social Care Act 2012), each CCG is duty bound by legislation to provide medical care to the community it serves.

[13] B. Godman, A. Bucsics, P.V. Bonanno, W. Oortwijn, C. Rothe, A. Ferrario, et al, "Barriers for Access to New Medicines: Searching for the Balance Between Rising Costs and Limited Budgets," *Frontiers in Public Health* 6 (2018).

[14] D. Howdon, N. Rice, "Health Care Expenditures, Age, Proximity to Death and Morbidity: Implications for an Ageing Population," *Journal of Health Economics* 57 (2018): 60-74.

[15] For a detailed discussion on the wider public spending on children's services in the UK, see: E. Kelly, T. Lee, L. Sibieta, T. Waters, *Public Spending on Children in England: 2000 to 2020* (London: Children's Commissioner Institute for Fiscal Studies, 2018).

ning may raise concerns about the just allocation of resources.[16] In addition, discussions relating to these concerns are further complicated by the imperative for the NHS to deliver chronic disease management and long-term care – something that the NHS was not designed to deliver at the outset.[17] Therefore, for a healthcare system to provide both emergency care *and* long-term chronic disease management, *and* to keep afloat within a financial budget, choices based on resource allocation and healthcare prioritisation are inexorable.[18]

Although not tangibly felt by patients engaging with NHS care, all treatments will have a price or at the very least a cost to it. How much money is available and how this money is distributed across the healthcare sector contributes to the notion of limited resources. As patient demand continues to outstrip healthcare delivery, tensions may develop where patients compete with one another for access to appropriate medical care. Hence, healthcare prioritisation is a necessity to handle patient waiting lists as demands continue to grow.[19]

There are various reasons that may explain this mismatch between demand and supply. Improvements and advances in medicine allow healthcare professionals to treat more diseases with increasing complexity. Treating patients with complex clinical problems are more expensive than less complicated cases. Novel therapies made available by the pharmaceutical companies command high prices, primarily because these companies need to generate profit in return for their investment in research and development. Furthermore, as people are living longer; patients expect to live longer. Thus, the nature of treatment would change over time and with it carries increased costs to the healthcare sector. Resources are indefinitely limited and the scarcity of resources can either affect the provision of treatment to all patients (*not enough*

[16] C. Sung, J. L. Herbst, "The Ethics of Caring for Hospital-Dependent Patients," *BMC Medical Ethics* 18 (2017): 75-80.

[17] A. Comas-Herrera, R. Wittenberg, L. Pickard, "Long-term Care for Older People in the United Kingdom: Structure and Challenges," in *Long-Term Care: Matching Resources and Needs*, ed. D. Challis (London: Routledge, 2004).

[18] A. Charlesworth, K. Bloor, "70 years of NHS Funding: How do we Know How Much is Enough?" *British Medical Journal* 361 (2018): k2373.

[19] S. Coughlin, D. Roberts, K. O'Neil, P. Brooks, "Looking to Tomorrow's Healthcare Today: A Participatory Health Perspective," *Internal Medicine Journal* 48:1(2018): 92-96.

resources for everyone) or the provision of treatment to all patients at a specific point in time (*not all patients can be treated now*).[20] Resource allocation also poses a dilemma: to what extent health prioritisation and rationing can be exercised without compromising the professional and ethical standards that govern the conduct of clinicians? Which category of patients should remain in acute healthcare facilities or be transferred to long-term care institutions to make way for patients who have a greater need for resource-intensive treatment? These are discourses that are heavily laden with resource allocation.

Healthcare prioritisation and rationing cannot be easily defined. The General Medical Council professional guidance states that allocative decisions affecting patients should be: "fair, based upon clinical need, and on likely effectiveness of treatment."[21] It also acknowledges that "treatment options available to patients may be affected by limits in resources."[22] This makes medical resource allocation a complex issue as patients' demands for treatment may fluctuate over time. One important feature of NHS philosophy is that patient access to treatment must be unrestricted, in other words, "equal access for equal need."[23] This aspiration is perhaps flawed, given that it is dependent on the presumption that the healthcare system is financed in an unlimited manner.

Resource-intensive medical treatments place significant financial burdens on the NHS. Typically, physicians routinely provide resource- and labour-intensive treatment for patients at an intensive care unit. Maximising the value of critical care and life-sustaining treatments by improving patient outcomes is vital in effective resource allocation.[24] Likewise, for critically ill infants like Charlie Gard and Alfie Evans, the neonatal intensive care unit (NICU) delivers the necessary treatments

[20] L. M. Fleck, "Controlling Healthcare Costs – Just Cost Effectiveness or 'Just' Cost Effectiveness?" *Cambridge Quarterly of Healthcare Ethics* 27 (2018): 271-283.

[21] General Medical Council, *Leadership and Management for All Doctors* (Manchester: General Medical Council, 2012): para. 85 (3).

[22] Ibidem, para. 84.

[23] R. Klein, "The National Health Service (NHS) at 70: Bevan's Double-Edged Legacy," *Health Economics, Policy and Law* 14:1 (2018): 1-10.

[24] N. A. Al-Saad, C. Skedgel, J. Nortje, "Principles of Resource Allocation in Critical Care," *BJA Education* 17:12 (2017): 390-395.

for children with life threatening clinical conditions. Previous studies on critically ill neonates demonstrated that the costs of life-sustaining treatment and the long-term care of such infants surviving intensive care were significant.[25] A recent economic analysis on the costs of neonatal intensive care revealed that the likely cost per life saved on a typical NICU was approximately £420,000.[26]

With the pressure to ensure that costly treatments are provided cost-effectively, the UK government established the National Institute for Health and Clinical Excellence (NICE).[27] The institution was originally designed to mitigate the consequences of regional health authorities exercising their powers on the availability of expensive treatments.[28] Largely, its remit is to disseminate appraisal reports on new medical technologies, therapies, and procedures that have a strong cost-effective clinical evidence base. NICE routinely uses "Quality Adjusted Life Year" (QALY) as a health economic tool in its analyses to deter-

---

[25] L. J. Mangham, S. Petrou, L. W. Doyle, E. S. Draper, N. Marlow, "The Cost of Preterm Birth Throughout Childhood in England and Wales," *Pediatrics* 123:2 (2009): 312-327. See also S. Petrou, T. Sach, L. Davidson, "The Long-Term Costs of Preterm Birth and Low Birth Weight: Results of a Systematic Review," *Child: Care, Health and Development* 27:2 (2001): 97-115.

[26] S. Watson, W. Arulampalam, S. Petrou, "The Effect of Health Care Expenditure on Patient Outcomes: Evidence from English Neonatal Care," *Health Economics* 26:12 (2017): e274-84.

[27] Originally set up as the National Institute for Clinical Excellence in 1999, the organisation merged with the Health Development Agency in April 2005 to form NICE. It is considered as an Executive Non-Departmental Public Body and its output is not an integral part of any UK governmental department. The institute is overseen by a board and members of that board are directly appointed by government ministers. NICE employs its own staff and enjoys an allocated budget from central government.

[28] NICE was established in the attempt to address the "postcode lottery" of healthcare services. Broadly, the term relates to the perceived unequal provision of services around England and Wales. The phenomenon is likely to be due to local budgetary constraints and decisions-making rather than differences in postcodes. N. Timmins, M. Rawlins, J. Appleby, *A Terrible Beauty: A Short History of NICE the National Institute for Health and Care Excellence* (Nonthaburi: Amarin, 2016).

mine whether a particular treatment is cost effective;[29] though alternate assessments exist.[30] This is a measurement that counts the cost of healthcare relative to its benefit on life expectancy. The cost per QALY is a rather important determinant in NICE recommendations; though how it uses such data in its decision-making process has been criticised as lacking a systematic approach.[31] In the UK, any treatment or intervention costing more than £30,000 per QALY is not normally provided by the NHS.[32] Furthermore, any endorsements made by the institute in its appraisal reports and clinical guidelines are generally binding on the purchasers of healthcare rather than on the clinicians.

Given that the issues surrounding resource allocation are pertinent to healthcare settings where resources are limited, this chapter will make an attempt to contribute to the discourse on the impact of resource allocation. The scope of this chapter will primarily focus on the provision of costly life-sustaining treatments in a resource-finite healthcare environment. Using the high-profile cases of Charlie Gard and Alfie Evans, this chapter discusses the clinical and legal issues illustrated by these cases from a resource allocation perspective. In addition, as the families of Charlie and Alfie had resorted to medical crowdfunding arguing that limited resources may have affected the care their sons received, a discussion on this relatively novel phenomenon is included within the context of resource allocation. This chapter will hold that although the notion of crowdfunding can be conceived as being well-intentioned, the practice of crowdfunding pressurises healthcare institutions and professionals to provide treatments which might not be in the best interests of their patients.

---

[29] For a detailed discussion on QALYs see: J. McKie, P. Singer, J. Richardson, *The Allocation of Health Care Resources* (London: Routledge, 1998), 20.

[30] A. Beresniak, D. Dupont, "Is There an Alternative to Quality-Adjusted Life Years for Supporting Healthcare Decision-making?" *Expert Review of Pharmacoeconomics & Outcomes Research* 16:3 (2016): 351-357.

[31] H. Dakin, A. Gray, "Decision-making for Healthcare Resource Allocation: Joint v. Separate Decisions on Interacting Interventions," *Medical Decision-making* 38:4 (2018): 476-486.

[32] I. Cleemput, M. Neyt, N. Thiry, C. De Laet, M. Leys, "Using Threshold Values for Cost per Quality-Adjusted Life-Year Gained in Healthcare Decisions," *International Journal of Technology Assessment in Health Care* 27 (2011): 71-76.

II. Resource allocation in the context of infant life-sustaining treatment

With improved rates of infant mortality made possible by advances in neonatal medicine throughout the 20th century,[33] clinicians have become victims of their own success. Infants born with severe incurable and life-limiting conditions are often managed on NICU from birth, where life and death medical decisions are made. With better rates in survival and higher expectations of what NICU can achieve, it is understandable that parents of critically ill children would advocate for continued life-sustaining treatments without being acutely aware of the costs involved. Kuhse and Singer pithily emphasised that in relation to the care of critically ill children on life-sustaining treatments:

> Where the chances of disability to an infant are small, the benefit should outweigh the associated costs. On the contrary, if the risks of severe disability are high and that there is the chance of the parents having a second pregnancy producing a healthy child, then the costs of NICU treatment may produce a loss rather than a gain.[34]

The cases of Charlie Gard and Alife Evans raise many clinical, ethical, and legal discussions on how they should be best managed by their respective clinical teams. Such a discussion would not be complete without an analysis of the associated costs and the resources that are necessary to provide life-sustaining treatment for these children. Recognising that life-sustaining treatment can be an emotive subject matter, resource allocation may offer a degree of objectivity on how realistic it is for children in a life-threatening condition to be kept alive on resource-intensive treatments.[35] Health economic studies have demonstrated that the application of objective cost-effectiveness analyses on infant intensive care may actually provide an ethically defensible ap-

---

[33] L. McKechnie, K. Johnson, "Quality of Neonatal Intensive Care and Outcome for High-Risk Newborn Infants," in *Neonatology,* ed. G. Buonocore, R. Bracci, M. Weindling (New York: Springer Cham, 2018).

[34] H. Kuhse, P. Singer, "Age and the Allocation of Medical Resources," *Journal of Medicine and Philosophy* 13:1(1988): 101-116, 112.

[35] N. Bhatia, *Critically Impaired Infants and End of Life Decision-making* (Abingdon: Routledge, 2017), 161-162.

proach to determine at what point resource-intensive life-sustaining treatments are potentially inappropriate.[36] Decisions on life-sustaining treatments that are simply based on cost per QALY raises an important conundrum – whether there are some infants who are simply too expensive to be treated in a healthcare setting where resources are finite.[37]

Finite healthcare resources dictate the availability of certain medicines, diagnostic equipment, bed capacity, and even healthcare professionals. Some patients may benefit from one therapy more than others; hence the most appropriate clinical treatment should be prioritised and rationed to those who would likely benefit the most. Healthcare prioritisation by its very nature, often results in the investment of resources to one part of the healthcare system whilst at the same time equally viable opportunities elsewhere are missed.[38] However, decisions on how resources be best spent cannot simply be based on "opportunity costs"; rather, such decisions are made with the relevant ethical, legal, clinical, and socio-economic values in mind. Where critical care is concerned, decisions can effectively be a matter of life and death.

In treating infants such as Charlie and Alfie, clinicians often jointly decide with parents on the likely course of treatment. In these circumstances, patient autonomy and informed consent are not applicable. Clinical decisions to provide life-sustaining treatments are mainly influenced by clinical need and suitability. Occasionally, healthcare professionals may be compelled to admit a child to NICU in order to facilitate parents to come to terms with the gravity of their child's illness or imminent death.[39] Providing aggressive and invasive treatment to save a

---

[36] D. Wilkinson, S. Petrou, J. Savulescu, "Rationing Potentially Inappropriate Treatment in Newborn Intensive Care in Developed Countries," *Seminars in Fetal and Neonatal Medicine* 23:1 (2018): 52-58.

[37] D. Wilkinson, "Which Newborn Infants are Too Expensive to Treat? Camosy and Rationing in Intensive Care," *Journal of Medical Ethics* 39 (2013): 502-506. See also, M. Pruski, "Experience Adjusted Life Years and Critical Medical Allocations within the British Context: Which Patient Should Live?" *Medicine, Health Care and Philosophy* 21:4 (2018): 561-568.

[38] D. A. Levinthal, "Resource Allocation and Firm Boundaries," *Journal of Management* 43:8 (2017): 2580–87.

[39] J. Boulais, T. Vente, M. Daley, S. Ramesh, J. McGuirl, B. Arzuaga, "Concern for Mortality in the Neonatal Intensive Care Unit (NICU): Parent and Physician Perspectives," *Journal of Perinatology* 38:6 (2018): 718-727.

critically impaired child from death by virtue of beneficence may therefore leave that child with severe lifelong disability, though life-sustaining treatments in such circumstances may not necessarily guarantee survival. Thus, the separation between acting in the child's best interests (beneficence) and not inducing harm (non-maleficence) whilst ensuring that all children are managed with the same compassion (distributive justice) is not always clear from the outset.

From a resource allocation perspective, devoting resources into expensive treatments to save a critically impaired infant with a poor prognosis may not necessarily produce benefit to society or be conducive to maximal utility of available resources.[40] As intensive care consumes a significant portion of resources, sustaining such treatment may limit other patients from treatment. Some commentators believe that this may in effect be an injustice to those being deprived of medical treatment.[41] Charlie's parents, having learned that nucleoside therapy may potentially benefit his genetic condition, argued that he should be provided with this expensive therapy. Although nucleoside therapy was not considered standard treatment for Charlie's condition, his doctors at the Great Ormond Street Hospital conditionally agreed to try this therapy on him.[42]

For both Charlie and Alfie, their doctors eventually realised that despite the life-sustaining treatments, both continued to clinically deteriorate and that their continued treatment was becoming increasingly futile.[43] Both teams of physicians proceeded to have emotionally difficult conversations with Charlie's and Alfie's parents commending a view that life-sustaining treatment be withdrawn, and palliative care be instituted. Understandably, both parents at that point demanded that current treatments be continued. In Charlie's case, his parents requested

---

[40] S. Gallagher, M. Little, C. Hooker, "The Values and Ethical Commitments of Doctors Engaging in Macroallocation: A Qualitative and Evaluative Analysis," *BMC Med Ethics* 19:1 (2018): 75.

[41] A. Puyol, "The Uncomfortable Truths of Healthcare Prioritisation," in *Prioritising Health Services or Muddling Through*, ed. J. E. del Llano-Señarís, S. Peiró-Moreno (Madrid: Springer Healthcare, 2018), 125-143, 126.

[42] *Yates v Great Ormond Street Hospital for Children NHS Foundation Trust* [2017] EWCA Civ 410 para. 20.

[43] Ibidem, para. 21; *Evans v Alder Hey Children's NHS Foundation Trust* [2018] EWCA Civ 805 para. 10.

that his doctors expedite his experimental nucleoside therapy despite rapid deterioration.[44] His doctors believed that such therapy was no longer feasible nor clinically appropriate given that the chance of survival was minute. Charlie's parents simply did not agree with that view.

Similarly, doctors at Alder Hey who cared for Alfie since the age of 7 months also concluded that his prognosis was becoming increasingly poor.[45] Recommendations of withdrawal of life-sustaining treatment to make way for palliative care were rejected by his parents. Alfie's family argued that he should be transferred to the Bambino Gesù Hospital in Rome for his ventilator support having sought assurances that Alfie would undergo a tracheostomy and a feeding tube for artificial nutrition and hydration.[46] Both families subsequently embarked on a bitter legal dispute with medical professionals.

Whilst professional codes of practice affirm that physicians are not obliged to provide treatments demanded by patients or to keep their patients healthy, any rights of the patient (or parents) to treatment are likely to be limited.[47] The General Medical Council (GMC) advises that physicians may offer to their patients "treatment that are clinically appropriate and of greatest benefit to which the patient can either accept or decline."[48] Moreover, the GMC also states: "you should not withdraw or decide not to start treatment if doing so would involve significant risk for the patient and the only justification is resource constraints."[49] The British Medical Association goes further advising that if

---

[44] Yates (n 42) para. 30.

[45] *Re E (A Child)* [2018] EWCA Civ 550 paragraphs 11-19.

[46] Ibidem, paragraphs 21 and 38.

[47] K. Dew, L. Signal, J. Stairmand, A. Simpson, D. Sarfati, "Cancer Care Decision-making and Treatment Consent: An Observational Study of Patients' and Clinicians' Rights," *Journal of Sociology* 55:1 (2019): 161-178. See also, K. Syrett, "Evolving the Right to Health: Rethinking the Normative Response to Problems of Judicialization," *Health Human Rights* 20:1 (2018): 121-132.

[48] General Medical Council, *Consent: Patients and Doctors Making Decisions Together* (Manchester: General Medical Council, 2008), https://www.gmc-uk.org/ethical-guidance/ethical-guidance-for-doctors/consent/part-2-making-decisions-about-investigations-and-treatment (accessed November 4, 2018).

[49] General Medical Council, *Treatment and Care Towards the End of Life: Good Practice in Decision-making* (Manchester: General Medical Council, 2010), para 39.

patients themselves feel benefit in receiving treatments deemed inappropriate, futile, or inadvisable, then effort should be made to honour these requests.[50]

In English case law, doctors are not legally bound to provide treatments that are considered clinically inappropriate or experimental irrespective of demand.[51] Some commentators have stated that doctors are acting ethically if: (i) they do not provide treatment judged inappropriate, futile, or contrary to the best interests of the patient;[52] and (ii) they do not provide treatment where resources are not available for this.[53] If Charlie were to receive experimental treatments abroad, and likewise Alfie was transferred to Rome for his treatments, perhaps this would release resources on intensive care so that they are made available to other children who require them. This argument, however, did not feature in the courts' deliberations for both children. Presumably, both cases fell on welfare grounds rather than on the interests of the parents or their rights to seek experimental or innovative treatments.[54]

The reasons that motivate patients or parental demands for treatment that are contrary to medical opinion are complex.[55] These are likely to be highly personal and are of specific importance to the individual and their families. As such, reasons underlying demands for inappropriate or futile treatments should be considered sensitively. Responsible clinicians, though they may rely on clinical guidelines and evidence-based practice to facilitate their decisions, should also consider the likely views of families. Decisions to delay withdrawal of life-

https://www.gmc-uk.org/-/media/documents/Treatment_and_care_towards_the_end_of_life___English_1015.pdf_48902105.pdf (accessed November 4, 2018).

[50] British Medical Association, *End-of-life Decisions - Views of the BMA* (London: BMA, 2006).

[51] *Re J (A Minor) (Wardship: Medical Treatment)* [1990] 3 All ER 930.

[52] T. M. Pope, "Legal Duties of Clinicians When Terminally Ill Patients with Cancer or Their Surrogates Insist on 'Futile' Treatment," *ASCO Post* (March 2018).

[53] D. Wilkinson, S. Petrou, J. Savulescu, "Expensive Care? Resource-Based Thresholds for Potentially Inappropriate Treatment in Intensive Care," *Monash Bioethics Review* 35 (2018): 2-23.

[54] G. Birchley, "Charlie Gard and the Weight of Parental Rights to Seek Experimental Treatment," *Journal of Medical Ethics* 44 (2018): 448-452.

[55] Ibidem.

sustaining treatment whilst the court formulates its judgement are not unusual.[56] Such considerate decisions are likely to be motivated by the interests of Charlie's and Alfie's parents. Whether the decision to withdraw treatment is taken in the courtroom or at the bedside, it is the case that they are likely to be challenging.[57] Medical resource allocation affects and constrains the ability of healthcare professionals to exercise "ideal" clinical practice whilst simultaneously be fair to all patients. Limited resources are incompatible with infinite availability and access to high cost treatments. Resources invested to treat one patient may preclude another to treatments enabled by those very same resources. Therefore, it is unsurprising that some clinical decisions have been challenged in the courtroom on the grounds of healthcare rationing and prioritisation. Likely precipitants for such legal disputes may include disagreements between clinicians and the patient (or parents) concerning treatment;[58] and the decision-making processes relating to medical treatment.[59]

Where a legal dispute occurs involving a child receiving life-sustaining treatment, the court would seek to determine whether a consensus exists between the clinicians and parents on the course of treatment.[60] This shared responsibility is subject to two presumptions. Firstly, parents must make decisions for their child based on the child's best interests not according to parental agenda. Secondly, medical professionals must exercise good medical practice by involving parents in their decision-making processes, but they have no duty to provide treatments that are against the child's best interests or are deemed clini-

[56] G. Birchley, R. Huxtable, "Critical Decisions for Critically Ill Infants: Principles, Processes, Problems," in *Pioneering Healthcare Law: Essays in Honour of the Work of Margaret Brazier*, ed. C. Stanton, T. Hervey, A. Farrell (London: Routledge, 2016).

[57] K. E. Miller, R. D. Coleman, L. Eisenberg, J. Henriksen, J. D. Lantos, E. M. Weiss, "Unilateral Withdrawal of Life-sustaining Therapy in a Severely Impaired Child," *Pediatrics* 142:5 (2018): e20180131.

[58] If clinicians and a child's family are in fundamental disagreement over the child's treatment, the views of the court should be sought. *Glass v the United Kingdom*, app no 61827/00, 39 EHRR 15 (2004).

[59] *Re B* [1981] 1 WLR 421; *Re C (A Minor)* [1989] 2 All ER 782; *Re J (A Minor) (Wardship: Medical Treatment)* [1990] 3 All ER 930; *Re* R *(Adult: Medical Treatment)* [1996] 2 FLR 99.

[60] *Re J (A Minor) (Child in Care: Medical Treatment)* [1992] 2 All ER 614.

cally inappropriate.[61] Thus, parents cannot coerce clinicians to provide treatment that is not medically indicated.

Whilst most disagreements on life-sustaining treatment may eventually be settled through dialogue between clinicians and parents outside the courtroom,[62] breakdown of trust between the parties may lead to a protracted dispute.[63] In both Charlie's and Alfie's cases, it has been suggested that their parents suffered from irreparable relationship breakdown with medical professionals.[64] Both families rejected medical recommendations presented to them preferring to settle the dispute legally. Through traditional and social media, both disputes subsequently became high profile and were widely publicised. In addition, Charlie's and Alfie's parents resorted to medical crowdfunding to gather support and perhaps social solidarity for their ongoing legal dispute. Though the parents' intentions were to raise funds and seek the necessary treatment they believed Charlie's and Alfie's doctors were unwilling to provide, it is suggestive that the parental actions implied that resource allocation *might* be a factor in the decision for the doctors to recommend withdrawal of treatment.

In scenarios where a dispute involves decisions regarding life-sustaining treatment for a child, judicial determination is often sought to answer whether it is in the child's best interests to receive treatment and whether it is in a child's best interests for the hospital to provide it.[65] A judicial inquiry of this nature not only considers the many relevant medical, emotional, and welfare factors, but regard would be given

---

[61] *An NHS Trust v S & Ors* [2003] EWHC 365 (Fam).

[62] J. P. Winters, "When Parents Refuse: Resolving Entrenched Disagreements Between Parents and Clinicians in Situations of Uncertainty and Complexity," *American Journal of Bioethics* 18:8 (2018): 20-31.

[63] A. Gallagher, "What Can We Learn from the Case of Charlie Gard? Perspectives from an Inter-Disciplinary Panel Discussion," *Nursing Ethics* 24:7 (2017): 775-777.

[64] E. Waldman, J. Frader, "Charlie Gard: How Did Things Go Wrong?" *Current Paediatric Reports* 6:2 (2018): 173-177.

[65] A. C. Elias-Jones, J. Samanta, "The Implications of the David Glass Case for Future Clinical Practice in the UK," *Archives of Disease in Childhood* 90:8 (2005): 822-825.

to the effects of limited resources on available treatment options.[66] Claims for judicial review may be brought if patients (or their family) believe that medical resource allocative decisions affecting them were illegal or irrational. They could also be brought on grounds of procedural impropriety.[67]

Based on existing case law, the court affirms that doctors may refuse to provide a medical treatment if it can be demonstrated that the treatment is medically contraindicated.[68] Courts recognise that they cannot compel doctors to pursue a particular course of treatment on a patient, but they can determine whether or not treatment would be in the patient's best interests.[69] On the other hand, courts also recognise the right of parents to advance their views on the welfare of their child.[70] Any treatment decisions must be fair, transparent, and not arbitrary.[71] Relevant clinical evidence and guidelines that were used to facilitate treatment decisions should be made available for scrutiny.[72]

In circumstances where parents make demands on the CCG to finance treatment for their children that are neither cost-effective nor clinically appropriate, concerns are often raised as to the obligations owed to the child and to the parents. In the widely publicised case of *R v Cambridge Health Authority ex parte B,*[73] which concerned a 10-year-old

---

[66] J. Hewitt, "Just Healthcare and Human Flourishing: Why Resource Allocation Is Not Just Enough," *Nursing Ethics* 26:2 (2019):405-417.

[67] A. Ford, "Accountability for Reasonableness: The Relevance, or Not, of Exceptionality in Resource Allocation," *Medicine, Health Care and Philosophy* 18 (2015): 217.

[68] *Re J (A Minor) (Child in Care: Medical Treatment)* [1992] 2 All ER 614; *Burke v GMC* [2005] EWCA Civ 1003.

[69] *Re A (Male Sterilisation)* [2000] FCR 193; and *Re S (Adult: Sterilisation)* [2000] 2 FLR 389. *Health & Social Services Trust v PM & Anor* [2007] NI Fam 13 (21 December 2007).

[70] E. Close, L. Willmott, B. P. White, "Charlie Gard: In Defence of the Law," *Journal of Medical Ethics* 44 (2018): 476-480.

[71] *North West Lancashire Health Authority v A, D & G* [1999] EWCA Civ 2022 [2000] 1 WLR 977.

[72] V. Larcher, F. Craig, K. Bhogal, D. Wilkinson, J. Brierley, "Making Decisions to Limit Treatment in Life-Limiting and Life-Threatening Conditions in Children: A Framework for Practice," *Archives of Disease in Childhood* 100 (2015): s1-s23.

[73] *R v Cambridge Health Authority ex parte B* [1995] 1 WLR 898.

child with acute leukaemia, the child's father sought to overturn the decision of refusal from the health authority to fund further chemotherapy and a bone marrow transplant. The health authority believed that any benefits of further chemotherapy and transplant were unproven and that the costs had greatly outweighed all perceived benefits. In the ruling, Laws J stated that decision-makers must "do more than to toll the bell of tight resources" and that the health authority must clarify their decision-making processes that had led them to refuse treatment. This judgement was subsequently overturned at the Court of Appeal and that "courts should concentrate on the lawfulness of the decision rather than on its merits." Sir Thomas Bingham MR develops this further:

> I have no doubt that in a perfect world any treatment which a patient or a patient's family sought would be provided if doctors were willing to give it, no matter how much it cost, particularly when a life was potentially at stake. It would however, in my view, be shutting one's eyes to the real world if the court were to proceed on the basis that we do live in such a world. It is common knowledge that health authorities of all kinds are constantly pressed to make ends meet. They [health authorities] cannot pay their nurses as much as they would like; they cannot provide all the treatments they would like; they cannot purchase all the extremely expensive medical equipment they would like; they cannot carry out all the research they would like; they cannot build all the hospitals and specialist units they would like. Difficult and agonising judgments have to be made as to how a limited budget is best allocated to the maximum advantage of the maximum number of patients. This is not a judgment which the court can make. In my judgment, it is not something that a health authority such as this authority can be fairly criticised for not advancing before the court.[74]

The Court of Appeal judgment had determined that the role of the court was neither to investigate actual decisions nor to consider matters about how limited resources are spent. The court ruled that its role was to scrutinise whether the decision-making processes were legal and

[74] R *v Cambridge Health Authority* [1995] EWCA Civ 43.

rational. To aid in that scrutiny, courts may consider whether those making the decisions had the power to do so, whether appropriate and correct procedures were followed, the relevant patient and clinical factors, and whether those decisions were proportionate. Courts are likely to regard clinicians to be acting reasonably if their actions or decisions were based on established clinical guidelines that stood up to logical analysis and on evaluations of likely cost-effectiveness and benefits.[75]

Traditionally, courts have argued that they lacked the clinical expertise to adjudicate on disputes about allocative decisions.[76] Many judges have expressed that decisions regarding treatment should be well left to those who have the necessary competence.[77] This has led to commentators suggesting that the court's over-reliance on medical opinion may affect its impartiality when presiding over disputes between patients and medical professionals over matters involving limited resources.[78] Moreover, judges are likely to apply terms such as "futility, best interests, quality of life, and suffering" when in reality the treatment is unlikely to yield any substantial benefit.[79] In the case of *Re J (A Minor) (Wardship: Medical Treatment)*,[80] Lord Donaldson made a cursory comment on finite resources, being mindful not to engage with resource allocation as it applied to the treatment of children:

> [I]t is outwith the scope of this judgment to give any guidance as to the considerations which should determine such an allocation…

[75] In *Re C (A Baby)* [1996] 2 FCR 569, "[t]he courts do not instruct doctors how they should perform their clinical and professional duties."

[76] K. Syrett, "Of Resources, Rationality and Rights: Emerging Trends in the Judicial Review of Allocative Decisions," *Web Journal of Current Legal Issues* (2000).
http://www.bailii.org/uk/other/journals/WebJCLI/2000/issue1/syrett1.html (accessed December 5, 2018).

[77] In *R v Central Birmingham Health Authority, ex parte Collier* (1988): "It is not for this court, or indeed any court, to substitute its own judgment for the judgment of those who are responsible for the allocation of resources."

[78] K. Syrett, "Courts, Expertise and Resource Allocation: Is There a Judicial 'Legitimacy Problem'?" *Public Health Ethics* 7:2 (2014): 112–122.

[79] N. Bhatia, *Critically Impaired Infants and End of Life Decision-making* (Abingdon: Routledge, 2017), 198.

[80] [1991] 1 Fam. 33.

In the seminal case of *Airedale NHS Trust v Bland,*[81] the court was likewise very reluctant to enter into any discourse on resource allocation. In the attempt to avoid discussion on the matter, Hoffman LJ stated:

> The resources of the National Health Service are not limitless and choices have to be made. This qualification on moral duty to provide care did not enter into the argument in this case at all. The Airedale NHS Trust invited us to decide the case on the assumption that its resources were unlimited and we have done so. But one is bound to observe that the cost of keeping a patient like Anthony Bland alive is very considerable and that in another case the health authority might conclude that its resources were better devoted to other patients. We do not have to consider such a case, but in principle the allocation of resources between patients is a matter for the health authority and not for the courts.

Careful not to guide healthcare professionals on how finite resources should be spent, Lord Browne-Wilkinson in *Bland* gave his perspective:

> In addition to these ethical questions, the new technology raises practical problems. Given that there are limited resources available for medical care, is it right to devote money to sustaining the lives of those who are, and always will be, unaware of their own existence rather than to treating those who, in a real sense, can be benefited, e.g. those deprived of dialysis for want of resources.

Though it is correct to assume that the NHS has a statutory duty to provide medical treatment,[82] courts appear to appreciate that with finite resources, there cannot be an expectation or duty to provide all conceivable treatments. In R *(on the Application of Coughlan & Ors) v North & East Devon Health Authority*, the court recognised that the Secretary of State for Health though he had a duty to promote a comprehensive healthcare service, he is also entitled to consider the amount of re-

[81] [1993] UKHL 17.
[82] National Health Service Act 2006 section 1(1b).

sources available to him and the demands placed on those services under his charge.[83] Furthermore, in R *v Secretary of State for Social Services*, the Court of Appeal determined that there is no absolute duty to provide treatment irrespective of economic decisions taken at the national level.[84]

Legal commentators have observed that the courtroom is not an appropriate place to resolve disputes that are primarily of a clinical nature.[85] Others have argued that courts merely add to unnecessary delays and costs to already difficult decision-making process[86] and that medical mediation may be a more appropriate avenue for dispute resolution.[87] The legal approach taken by the court in relation to the cases of Gard and Evans identifies three observations that can perpetuate the dispute between the parents of these children and the medical professionals. Firstly, there is inertia of the court to engage with the impact that limited resources may exert on treatment decisions. Secondly, there is a lack of a standard best interests test for children which are unlike that of adults without mental capacity.[88] Thirdly, there is the perception of over-reliance of the court on medical opinion and the insufficient weighting it affords to parental rights to advocate on behalf of their child's best interests. A combination of these factors may have contributed to the unsatisfactory legal outcome for both families. Charlie and Alfie did not go on to receive the treatments so sought after by their parents.

---

[83] R *(On the Application of Coughlan & Ors) v North & East Devon Health Authority* [1999] EWCA Civ 1871 para. 26.

[84] R *v Secretary of State for Social Services, ex parte Hincks* [1980] 1 B.M.L.R. 93.

[85] T. M. Pope, "Guest Editorial: Charlie Gard's Five Months in Court: Better Dispute Resolution Mechanisms for Medical Futility Disputes," *Journal of Medical Ethics* 44:7 (2018): 436-437.

[86] R. Huxtable, "Clinic, Courtroom or (Specialist) Committee: In the Best Interests of the Critically Ill Child?" *Journal of Medical Ethics* 44:7 (2018): 471-475.

[87] S. Meller, S. Barclay, "Mediation: An Approach to Intractable Disputes Between Parents and Paediatricians," *Archives of Disease in Childhood* 96:7 (2011): 619-621.

[88] D. Wilkinson, S. Barclay, J. Savulescu, "Disagreement, Mediation, Arbitration: Resolving Disputes About Medical Treatment," *Lancet* 391:10137 (2018): 2302-2305.

One remarkable similarity in the cases of Charlie and Alfie was their successful social media and crowdfunding campaigns. The global attention on the plight for both infants attracted comments from prominent individuals such as the comments from the President of the United States and that of Pope Francis.[89] Both families had raised significant sums to finance treatments for their children having learned that their respective hospital institutions were actively considering withdrawal of life-sustaining treatment. Though the families cannot be criticised for their fundraising endeavours, the next section will discuss the implications of crowdfunding on resource allocation. As a means of funding, medical crowdfunding cannot be a long-term solution to the problems associated with healthcare rationing and prioritisation.

III. The rise of crowdfunding for resource-intensive medical treatment

The cases of Charlie Gard and Alfie Evans are perhaps high profile examples of medical crowdfunding in the UK. The first cases of medical crowdfunding were reported in the United States[90] and its appeal has subsequently led to increasing global popularity as a fundraising method to finance expenses related to healthcare. In countries, for example in the UK, where access to healthcare is based on the premise of a "single" payer[91] (general taxation) to fund costs incurred in relation to diagnosis and treatment, medical crowdfunding may offer a means for patients to appeal for social solidarity and support.[92] This may not necessarily hold true in healthcare systems where access to healthcare is

---

[89] Samuelson, K. *What to Know About Charlie Gard, The Terminally Ill Baby Trump Wants to Help*. New York: Time Magazine, 2017. http://time.com/4843746/president-trump-pope-francis-charlie-gard/ (accessed November 30, 2018).

[90] L. S. Berliner, N. J. Kenworthy, "Producing a Worthy Illness: Personal Crowdfunding Amidst Financial Crisis," *Social Science and Medicine* 187 (2017): 233-2.

[91] J. Oberlander, "The Virtues and Vices of Single-Payer Health Care," *New England Journal of Medicine* 374 (2016): 1401-1403.

[92] S. Alijani, D. Assadi, A. Ashta, "Can Social Dynamics Be Channeled from Offline to Online Communities?: Commitment, Compliance, and Cooperation from Microfinance to Crowdfunding," in *Strategic Approaches to Successful Crowdfunding*, ed. D. Assadi (Hershey PA: IGI Global, 2016).

considered not as a shared responsibility, but rather a privilege or luxury.[93]

Charlie's parents began a fundraising campaign for nucleoside therapy after learning that such a therapy was not standard treatment at Great Ormond Street Hospital. To cover the costs for this experimental therapy, they turned to crowdfunding. When their request to transfer their child's care to the United States was denied at the High Court, Charlie's parents announced their intent to use the funds raised to establish The Charlie Gard Foundation with the aim to finance research into mitochondrial diseases.[94] In Alfie Evan's case, his parents also resorted to medical crowdfunding to establish The Alfie Evans Foundation with the aim to provide financial assistance for families with critically ill children with "undiagnosed" medical conditions.[95]

The Gard family's GoFundMe webpage was immensely successful. The fund amounted to an excess of £1.3 million since January 2017.[96] In comparison, the Evans family's JustGiving crowdfunding webpage raised in excess of £142,000 within six months by June 2018.[97] The Charlie Gard Foundation continues to receive crowdfunding donations, while the Alfie Evans crowdfunding webpage has ceased receiving donations since his tragic death. Although other medical crowdfunding projects have raised substantial sums, Charlie's parent's successful campaign was remarkable and beyond what is usual of typical crowdfunding achievements.[98]

Given the attraction of appealing to the public at large to share the burden of healthcare expenses, many commentators have drawn atten-

---

93 J. D. Lenchus, "The Right to Health Care," *Journal of the American Medical Association* 317:13 (2017): 1377-1378.

94 GoFundMe. Charlie Gard #charliesfight.
https://www.gofundme.com/please-help-to-save-charlies-life
(accessed December 4, 2018).

95 JustGiving. Alfie Evans #alfiesfight.
https://www.justgiving.com/crowdfunding/sarah-evans-791
(accessed December 4, 2018).

96 *R* (n 73).

97 Ibidem.

98 A. L. Gonzales, E. Y. Kwon, T. Lynch, N. Fritz, "Better Everyone Should Know Our Business than We Lose Our House: Costs and Benefits of Medical Crowdfunding for Support, Privacy, and Identity," *New Media and Society* 20:2 (2018): 641–58.

tion to the ethical issues of such practices.[99] Almost all crowdfunding campaigns begin with an extensive disclosure of private, medical, and confidential information in the form of a narrative to maximise appeal to potential donors and their donations. Both Charlie's and Alfie's parents had made a conscious decision to extensively reveal their child's personal and medical information online. Whilst this may raise ethical concerns in relation to confidentiality, nevertheless parents are legally permitted to disclose information about their children.[100]

Medical crowdfunding seemingly appears to benefit those who are vociferous, photogenic (for the purposes of the both online and traditional media), and the technologically resourceful. In addition, those who maintain a large online social network and receive endorsements from prominent figures are likely to have improved chances in running successful crowdfunding campaigns.[101] Factors such as perceived social value, emotional attachment, and donor sentimentality can heavily influence the fundraising outcome.[102] These run contrary to standard considerations like clinical need and benefit. There is currently a paucity of data in the medical literature to show how funds raised from crowdfunding are used in the UK though it is reasonable to suggest that individuals may utilise such funds to finance private or experimental treatments;[103] for example, the nucleoside therapy offered in the United States for Charlie's genetic condition. It is clear that access to treat-

---

[99] J. Snyder, "Crowdfunding for Medical Care: Ethical Issues in an Emerging Health Care Funding Practice," *Hastings Center Report* 46 (2016): 36-42. See also: J. Young, E. Scheinberg, "The Rise of Crowdfunding for Medical Care: Promises and Perils," *Journal of the American Medical Association* 317:16 (2017): 1623–1624.

[100] C. Bessant, "Sharenting: Balancing the Conflicting Rights of Parents and Children," *Communications Law* 23:1 (2018): 7-24.

[101] J. Snyder, V. A. Crooks, A. Mathers, P. Chow-White, "Appealing to the Crowd: Ethical Justifications in Canadian Medical Crowdfunding Campaigns," *Journal of Medical Ethics* 43:6 (2017): 364-367.

[102] G. Bassani, N. Marinelli, S. Vismara, "Crowdfunding in Healthcare," *Journal of Technology Transfer* (April 2018).

[103] M. J. Young, E. Scheinberg, "The Rise of Crowdfunding for Medical Care: Promises and Perils," *Journal of the American Medical Association* 317:16 (2017): 1623–1624.

ments financed by medical crowdfunding is likely to be non-egalitarian[104] in nature.[105]

Charlie's and Alfie's cases illustrate how medical crowdfunding can distort clinical decision-making processes in favour of patients with compelling or exceptional circumstances. In healthcare systems like the NHS, medical crowdfunding may potentially perpetuate the problems associated with healthcare rationing and prioritisation constrained by limited resources.[106] It may also diminish the need for quality improvement to redress problems with patient access to resource-intense treatments. Snyder et al assert that medical crowdfunding may also skew decision-making based on clinical need to that based on highly subjective parameters.[107] Moreover, the success of The Charlie Gard Foundation showed how medical crowdfunding can shift and divert limited resources to the treatment of rare and incurable genetic disorders over common and cheaper-to-treat childhood diseases. In practice, crowdfunding places undue pressure on decision-makers to shift resources to benefit those based on exceptionality over clinical necessity.[108] Therefore, crowdfunding as a concept compromises the value of medical advice relative to financial feasibility and creates interference with the physician's duty to act in the best interests of their patients. Whilst services delivered through the NHS are not subjected to market forces,

[104] Egalitarianism is the philosophical school of thought that prioritises equality for all individuals and egalitarian healthcare doctrines advocate that all patients should be treated on a similar basis. Conceivably, the opposite of egalitarianism is elitism; medical crowdfunding perhaps may grant some individuals better access to treatment based upon the ability to pay.

[105] M. J. Renwick, E. Mossialos, "Crowdfunding our Health: Economic Risks and Benefits," *Social Science and Medicine* 191 (2017): 48-56.

[106] G. Dressler, S. A. Kelly, "Ethical Implications of Medical Crowdfunding: The Case of Charlie Gard," *Journal of Medical Ethics* 44:7 (2018).

[107] K. Syrett, "Courts, Expertise and Resource Allocation: Is There a Judicial 'Legitimacy Problem'?" *Public Health Ethics* 7:2 (2014): 119.

[108] R. D. Truog, "The United Kingdom Sets Limits on Experimental Treatments: The Case of Charlie Gard," *Journal of the American Medical Association* 318:11 (2017): 1001-2. doi: 10.1001/jama.2017.10410.

crowdfunding may surreptitiously introduce market norms and how society perceives the notion of healthcare.[109]

Given the unproven benefits of nucleoside therapy, the Gard family intended to finance this treatment with little expectation on its likely efficacy. Some commentators have argued that patients should be free to buy any treatment they can afford even if it is against medical advice insofar as this action does not adversely impact on the healthcare delivery for other patients.[110] Even if funds realised through crowdfunding were used to buy private nucleoside treatment for Charlie, there would be an anticipated corresponding cost to the NHS.[111] Medical crowdfunding empowers patients (or in the case of Charlie and Alfie, their parents) to adopt a somewhat transactional relationship with medical professionals. This might engender disputes between parents who believe that they know best for their child and medical professionals who also have the child's best interests at heart. As the trend in the use of crowdfunding webpages to finance medical treatment continues to grow,[112] it is imperative that medical professionals appreciate the potential implications of this phenomenon on their clinical practice.

IV. Lessons learned from the cases of Charlie Gard and Alfie Evans in relation to resource allocation

The cases of Charlie Gard and Alfie Evans demonstrate that in relation to life-sustaining treatment, resource allocation is a valid consideration for healthcare professionals. Although the decision to fatally withdraw treatment for both infants cannot be wholly based on availability of enough resources, cost-effectiveness and clinical appropriateness should be balanced against the likely benefit of the therapy to the patient and best interests. The experience of Charlie's and Alfie's families

---

109 J. Snyder, P. Chow-White, V. A. Crooks, A. Mathers, "Widening the Gap: Additional Concerns with Crowdfunding in Health Care," *The Lancet Oncology* 18:5 (2017): e240.

110 R. Smith, C. Peedell, "Is the Private Sector a Good Thing for the NHS?" *British Medical Journal* 349 (2014): g5865.

111 *Re J (A Minor)(Wardship: Medical Treatment)* (n 80) p. 1002.

112 J. Snyder, A. Mathers, V. A. Crooks, "Fund My Treatment: A Call for Ethics-Focused Social Science Research into the Use of Crowdfunding for Medical Care," *Social Science and Medicine* 169 (2016): 27-30.

were painfully played out in the courtroom and this leads one to contemplate whether the legal system can effectively adjudicate disputes that are primarily of a clinical nature. With presiding judges seen to rely heavily on medical opinion, legitimate concerns are raised as to whether the court can be an impartial arbitrator between disputing parties on matters of limited resources.

Life-sustaining treatment is both resource- and labour-expensive. This requires and consumes significant resources and may not necessarily maximise utility of these resources to benefit all patients. Discussions about resource allocation may perhaps provide an objective approach to end-of-life decision-making for critically ill children like Charlie and Alfie. The families of both children have held, and will continue to hold, suspicions that treatment for their sons might be influenced by limited resources. This may explain why they had resorted to medical crowdfunding to raise the necessary funds so that treatments can be bought to prolong life as much as possible. This would feed into the psychological disposition to help individuals in imminent danger, especially if one feels that there is any possibility to prevent harm – a practice referred to as "Rule of Rescue."[113] With this frame of mind, decision-makers may simply divert limited resources to help one individual irrespective of the idea that many more could be saved by those same resources.

V. Conclusion

Resource-intense life-sustaining treatments are commonly provided to patients in life-threatening states. Although the NHS receives significant sums of money from central government with promised year-on-year increases,[114] novel medical advances along with the ability of medical professionals to prolong life continue to apply pressure on the resources at its disposal. In an ideal world where healthcare resources are infinite, rationing and prioritisation would simply be hypothetical arguments. However, the NHS is far from ideal. Patients cannot receive every treatment conceivable and priorities must be set to enable maxi-

[113] W. Lübbe, "Appeal to the Rule of Rescue in Health Care: Discriminating and Not Benevolent?" *Medicine, Health Care and Philosophy* 22:1 (2019): 53-58.
[114] C. Ham, "A Funding Boost for the NHS in England," *British Medical Journal* 361 (2018): k2741.

mal utility from these limited resources. Traditionally, clinicians decide which patients should receive treatment based on clinical need and likely benefit without consideration of the costs involved. However, this approach assumes that it is in the patients' best interests to receive such treatment and that the treatment itself will enable the patients to experience a better quality of life.

This chapter has discussed how resource allocation can impact decisions on life-sustaining treatment. The imperative for medical professionals to provide cost-effective treatment and the need to conduct themselves as fair to all patients under their care are affected by the resources made available to them. Legal action can have its uses – money to fund treatment may suddenly appear though this is likely to be due to diversion of existing resources in response to the threat of going to the courtroom. Whilst there have been legal cases where treatments have been refused to patients on grounds of limited resources, the courts have been rather reluctant to fully engage with the issue preferring to defer to medical opinion.[115] As a consequence, there is little evidence in case law that taking legal action against clinical decisions made by medical professionals would necessarily secure the desired treatment demanded by a patient (or parents). In a future where there are more patients with complex problems, increased availability of expensive medical therapies, and shifting expectations of healthcare, it can be anticipated that there would be more disputes referred to the law courts. Can the English legal system ever monitor and supervise healthcare resource allocation and judges be apposite arbitrators of disputes involving healthcare rationing? The answer might lie within the legal approach embraced by the court.[116] If courts continue to hold due process over efficiency, adjudication would be fair albeit rather slow-

[115] J. J. Paris, J. Ahluwalia, B. M. Cummings, M. P. Moreland, D. J. Wilkinson, "The Charlie Gard Case: British and American Approaches to Court Resolution of Disputes Over Medical Decisions," *Journal of Perinatology* 37:12 (2017): 1268–1271.

[116] C. Newdick, "Can Judges Ration with Compassion? A Priority-Setting Rights Matrix," *Health and Human Rights Journal* 20:1(2018): 107-120.

ly.[117] Perhaps more cases would help refine the court's role in settling disputes of this nature.

Given the popularity of social media, court actions such as that of Charlie Gard and Alfie Evans amplify the legal plight of both families.[118] The media may attract willing third-party donors to the parents' cause to crowdfund their sons' treatments that they believe are being denied. Medical crowdfunding potentially distorts decision-making in relation to clinical treatment to favour those with the ability to pay rather than that based on clinical need. Used in conjunction with the threat of legal action, it is possible that doctors and hospital authorities are compelled to defensively[119] provide treatments for their patients that are not clinically indicated or worse, futile.

[117] T. M. Pope, "Guest Editorial: Charlie Gard's Five Months in Court: Better Dispute Resolution Mechanisms for Medical Futility Disputes," *Journal of Medical Ethics* 44:7 (2018): 436-437.

[118] P. Schulz, A. J. Cannon, "Trial by Tweet? Findings on Facebook? Social Media Innovation or Degradation? The Future and Challenge of Change for Courts," *International Journal for Court Administration* 5:1 (2013).

[119] L. McCrossan, R. Siegmeth, "Demands and Requests for 'Inappropriate' or 'Inadvisable' Treatments at the End of Life: What Do You Do at 2 o'clock in the Morning When …?" *British Journal of Anaesthesia* 119:S1 (2017): i90-i98.

Chapter 6

Citizenship at the Discretion of the State: Public Law Issues Regarding Evans' Naturalisation

Alejandra Boto
University of Oviedo, Spain

I. Introduction

On April 24, 2018, many media outlets around the world announced that the Italian Ministry of Foreign Affairs had granted citizenship to Alfie Evans.[1] The move seemed to boost his parents' hope for a treat-

[1] Including, but not limited to: "Alfie Evans Granted Italian Citizenship After Pope Steps in as Family Hope Country Can Treat Toddler," https://www.telegraph.co.uk/news/2018/04/23/protesters-storm-hospital-terminally-ill-alfie-evans-parents/; "British Toddler Alfie Evans Not Allowed to Leave Country, UK Court says," https://edition.cnn.com/2018/04/25/health/alfie-evans-appeal-bn/index.html; "Royaume-Uni: Alfie Evans, 23 Mois, au Cœur d'un Débat sur la fin de vie" [United Kingdom: Alfie Evans, 23 months, at the Heart of an End-of-life Debate]. http://www.lefigaro.fr/international/2018/04/24/01003-20180424ARTFIG00147-alfie-evans-23-mois-au-centre-d-un-debat-sur-la-fin-de-vie.php; "Italia da la Nacionalidad al bebé Terminal que Reino Unido Quiere Desconectar" [Italy Confers Citizenship to the Terminally Ill Baby whom United Kingdom Wants to Unplug]. https://www.lavanguardia.com/vida/20180423/442909820740/italia-nacionalidad-alfie-evans-bebe-terminal-reino-unido.html; "L'Italie Octroie la Citoyenneté à Alfie Evans, Unbébé Britannique dans un état Semi-végétatif Menacé d'être débranché" [Italy Confers Citizenship to Alfie Evans, a British Baby in a Semi-Vegetative Status who Risks to be Disconnected], https://www.francetvinfo.fr/monde/italie/l-italie-octroie-la-citoyennete-a-alfie-evans-un-bebe-britannique-dans-un-etat-semi-vegetatif-menace-d-etre-debranche_2720096.html; "Il Cdm ha Conferito la Cittadinanza Italiana ad

ment abroad, involving at the same time important legal and diplomatic issues. Indeed, some media also echoed that the Italian ambassador was urgently to contact British courts in order to request Evans' immediate transfer to Italy.[2]

At the time, awarding Italian citizenship to someone who was born and resident in the United Kingdom and whose family had no Italian lineage was suspected to have a confessional dimension, mainly due to several public statements by Pope Francis which appeared in social networks.[3] Moreover, the move appeared to collide with some well-known principles of international law such as non-interventionism (i.e.

Alfie Evans" [The Council of Ministers has Conferred Italian Citizenship to Alfie Evans], http://www.ilgiornale.it/news/cronache/cdm-ha-conferito-cittadinanza-italiana-ad-alfie-evans-1519164.html; "Alfie Evans, Alfano e Minniti Concedono la Cittadinanza Italiana al Bambino. Lo Stop alle cure è Rinviato" [Alfie Evans, Alfano e Minniti have Conferred Italian Citizenship to the Baby. Cessation of Medical Attention is Postponed], https://www.ilfattoquotidiano.it/2018/04/23/alfie-evans-alfano-e-minniti-concedono-cittadinanza-italiana-e-il-distacco-delle-macchine-viene-sospeso-per-un-cavillo/4310491; "Der Kleine Alfie ist für Immer eEngeschlafen" [Little Alfie has Fallen Asleep for Good]. https://www.bild.de/news/ausland/krank/krankes-baby-stirbt-mit-23-monaten-55538934.bild.html; "El niño al que se Retiró el Soporte Vital en Reino Unido no Podrá Recibir Tratamiento en Italia" [The Child to whom United Kingdom Denied Care Would Not be Assisted in Italy]. https://elpais.com/internacional/2018/04/24/actualidad/1524589613_252635.html (all accessed November 12, 2018).

[2] For instance, "Alfie Evans Granted Italian Citizenship in Life Support Legal Row." https://www.bbc.com/news/uk-england-merseyside-43872678 (accessed November 11, 2018).

[3] The public offers made by the manager of the Bambino Gesù Hospital (Rome) to take care of the baby for "humanitarian" reasons and without charge also contributed to a religious-based discussion as the hospital itself is part of the Holy-See: F. d'Agostino, "Gli Sconfortati Dibattiti Sul Caso Alfie [The Discouraging Debate on the Alfie Evans Case]," *BioLaw Journal – Rivista di Biodiritto* 2 (2018): 23. However, besides the Catholic perspective, there was also a cultural involvement in the way Italy defended *civility* against British *barbarity*, as critically described by C. Barbisan, "Gli Sguardi Italiani su Alfie Evans, [Italian View on Alfie Evans]" *BioLaw Journal – Rivista di Biodiritto* 2 (2018): 18. Therefore it was no surprise that the citizenship was finally conferred by the Italian state and by the Vatican.

non-interference in domestic affairs) or effective nationality (which implies that the acquisition of nationality requires a genuine link between the individual and the state).

On the other hand, it is also perfectly known and a guiding principle of international relationships that it is for each state to determine under its own law who are its nationals. Thus Italian regulations allow an *ex gratia* conferral of citizenship. This method of acquiring nationality, which can be named as the privilege of extraordinary naturalisation,[4] is not subject then to the general rules of administrative procedure, although some formal requirements need to be verified. Naturalisation is a complex process, defined by configurations of policies and requirements that include material and procedural conditions,[5] as well as having an international law dimension.[6]

The toddler's death on April 28, 2018 suddenly closed the case and silently put an end to its practical derivations, but the theoretical framework and the potential implications are far from straightforward. They deserve a closer look, hereby approached from a public law perspective.

This chapter will first focus on the Italian regulation on this kind of naturalisation, which also exists in similar conditions in some other countries. Afterwards, it will deal with its implementation in the case of Alfie Evans in order to assess whether there was compliance between actual facts and legal provisions.

---

[4] See R. Clerici., *La Cittadinanza Nell'ordinamento Giuridico Italiano* [Citizenship in Italian Law] (Padova: Cedam, 1993), 342.

[5] S. Wallace Goodman, *Naturalisation Policies in Europe: Exploring Patterns of Inclusion and Exclusion* (Florence: EUDO Citizenship Observatory, 2010), 4.

[6] In sum, nationality rules are domestic matters only in so far as international law (including general principles, custom or international agreements) does not regulate the practice to the contrary, and as long as domestic rules do not otherwise conflict with international law: A. Edwards, "The Meaning of Nationality," in *Nationality and Statelessness under International Law*, eds. Alice Edwards and Laura van Waas (Cambridge: Cambridge University Press, 2014), 29.

II. The privilege of ultra-extraordinary naturalisation

Generally speaking, debates regarding the systems of citizenship conferral are dominated by the paramount dichotomy *jus sanguinis/jus soli*,[7] without forgetting two other widespread criteria, although sometimes considered minor: marital status and residence.[8] Another outstanding dichotomy is that of dividing the ways of acquiring citizenship among those by-birth and those after-birth (the latter usually known as naturalisation).[9]

Besides that, many states have a special loophole that allows a bypass regarding the ordinary methods of obtaining nationality. It is an extraordinary way of citizenship acquisition, often neglected as it is considered residual and discretionary.[10] Within that group, specialities regarding refugees, former nationals, stateless people, or sets with ethic/cultural ties may be outlined. Last but not least, a very particular detour in extraordinary naturalisation is the acquisition of citizenship by a full discretionary decision, reserved for individuals with special achievements for the country.

A comparative study carried out in 2006 identified this last kind of ultra-extraordinary naturalisation in some countries among the EU-15,[11] with a legal framework particularly detailed in the case of Spain.[12] Worldwide, however, it is not a unique case, as demonstrated by the

---

[7] *Jus sanguinis* is a Latin phrase that literally translates to "right of blood" in English; it asserts that an individual can claim a right to a citizenship according to the nationality or citizenship of a parent. *Jus soli* means "right of the soil" and it is the right of anyone to the citizenship of the state where they were born.

[8] P. Weil, "Nationalities and Citizenships," in *Citizenship, Nationality and Migration in Europe*, eds. David Cesarani and Mary Fulbrook (London: Routledge, 1996), 75-76.

[9] S. Tönsmann, *What is Citizenship for? Citizenships and Naturalization in Latvia* (Berlin: Berlines Wissenschafts-Verlag, 2017), 54.

[10] H. Waldrauch, "Acquisition of Nationality," in *Acquisition and Loss of Nationality*, eds. Rainer Bauböck, Eva Ersbøll, Kees Groendendijk, and Harald Waldrauch (Amsterdam: Amsterdam University Press, 2006), 176-177.

[11] Ibidem, 176.

[12] R. Rubio Marín, "Spain," in *Acquisition and Loss of Nationality*, ed. Rainer Bauböck, Eva Ersbøll, Kees Groendendijk, and Harald Waldrauch (Amsterdam: Amsterdam University Press, 2006), 477-478.

naturalisation of Bobby Fisher in Iceland, Donald Keene in Japan or, more recently, of several athletes in Qatar and Bahrain. In such cases, the state discretion, which is always present in an extraordinary naturalisation, is exacerbated by the use of undefined legal concepts and the lack of specific legal requirements such as the need to be domiciled in the country or to pass a language test.

Italy contemplates a version of such an ultra-extraordinary naturalisation in article 9.2, Act number 91, passed on February 5, 1992, which allows conferring citizenship by a discretionary government decision provided there is a special interest of the state. The first step in order to throw some light upon such a statute is to discuss the legal provision itself, which may be translated as follows:[13] *By Decree of the President of the Republic, having heard the Council of State and after consideration by the Council of Ministers, on a proposal of the Department in charge of Internal Affairs (Home Office), in consultation with Foreign Affairs, citizenship may be granted to an alien when they had made eminent services to Italy, or when there is an extraordinary interest of the state.*

An analysis of the text is provided below, focusing on four main aspects of the political decision involved: cause, nature, procedure, and effects. Afterwards, its application to the case of Alfie Evans will be discussed.

Spanish law, which is not only one of the most accurate regarding ultra-extraordinary naturalisation but is also, as will be demonstrated, very similar to the Italian regulation, is then considered for a comparative law approach.[14]

---

[13] In original: "Con decreto del Presidente della Repubblica, sentito il Consiglio di Stato e previa deliberazione del Consiglio dei Ministri, su proposta del Ministro dell'interno, di concerto con il Ministro degli affari esteri, la cittadinanza puo' essere concessa allo straniero quando questi abbia reso eminenti servizi all'Italia, ovvero quando ricorra un eccezionale interesse dello Stato."

[14] The precision of Spanish law was pointed in the EU-15 research referenced *supra* and as far as I know no other study has tagged a more relevant parameter for comparison. A comparative law perspective, even using an international tone, has been employed in recent studies carried out by Italian academics. Unfortunately, they tend to only focus on ordinary after-birth acquirement of citizenship and do not provide any help to the topic of this chapter. See, for instance, M. D'Orico, "Il riconoscimento dello status di cittadino alla luce di

A. Cause

The general assessment of this type of citizenship acquisition is unpromising, as it is built on imprecise legal remissions.[15] That is obviously the case regarding Italian regulation and the reference to "eminent services," as well as the concurrence of an "extraordinary interest" for the country. The phenomenon is even more acute in Spanish regulation, which allows the government discretionary prerogative to grant citizenship to an applicant who meets "exceptional circumstances."[16]

In order to analyse the potential range of those references and try to contemplate the kind of situations they might cover, the search for precedents is essential. Nevertheless, case law regarding this kind of citizenship acquisition is almost non-existent, both in Italy[17] and Spain.[18]

Furthermore, it is impossible in Italy to search for even non-binding precedents, as there is no official public list of citizenship conferrals under this statute. Some researchers do present several examples, but without a comprehensive examination or detailed information. For example, Professor Elisabetta Lamarque refers to several cases before

---

alcune esperienze internazionali" [Recognition of Citizen Status in Light of Some International Experiences], in *Una nuova cittadinanza. Per una riforma della legge del 1992* [A New Citizenship. For a change in 1992 Act], ed. Ennio Codini and Marina D'Odorico (Milan: FrancoAngeli, 2007), 71-90.

15 Nevertheless, there are also opinions supporting the potential of the privilege in order to correct, in particular cases, the injustices produced by a too rigid legal system for the acquisition of nationality. See for instance G. Cerdeira Bravo de Mansilla, "Nacionalidad por carta de Naturaleza: un Ejemplo de Equidad" [Citizenship by Extraordinary Naturalisation: An Example of Equity], *Anuario de Derecho Civil* [Civil Law Annual] LXIII (2010): 1219-1244.

16 Spanish Civil Code, article 21.1: "La nacionalidad española se adquiere por carta de naturaleza, otorgada discrecionalmente mediante Real Decreto, cuando en el interesado concurran circunstancias excepcionales."

17 P. Morozzo della Rocca, "La Naturalizzazione Dello Straniero" [Foreigners' Naturalisation], in *Immigrazione e Cittadinanza. Profili Normativi e Orientamienti Giurisprudenziali* [Immigration and Citizenship. Regulation Profile and Case Law Orientations], ed. Paolo Morozzo della Rocca (Turin: Utet Giuridica, 2008), 367-372.

18 A. Álvarez Rodríguez, *Nacionalidad Española. Normativa e Interpretación Jurisprudencial* [Spanish Citizenship. Regulation and Jurisprudential Interpretation] (Cizur Menor: Thomson-Aranzadi, 2008), 104-108.

Alfie Evans' since 2013, all related to sports merits, labour activity, or solidarity for tragic events linked to racism or cultural diversity in Italian territories.[19] That seems to be, for instance, the reason why Nosheen Ahmad But became Italian in 2015 after a family drama caused by her opposing an arranged marriage in Modena.[20] Via media news, one can highlight the conferral of citizenship for special merits in Italy to the musician Lazar Berman[21] or to Professor Joseph Halevi Hurwitz Weiler,[22] but an accurate and thorough inventory is unavailable.

Meanwhile, in Spain, it is easier to form an opinion, as the Royal Decrees conferring citizenship are published in the Official State Gazette. A simple search on the institutional website offers over 1500 examples.[23] Usually, someone meets "exceptional circumstances" and deserves to become Spanish because of military, political, economic, cultural, or sport merits or when one is in a defenceless situation.[24] Researchers such as Professor Aurelia Álvarez Rodríguez provide statistics for reasons ranging from royal lineage, diplomacy, to securing peace. The cast of privilege holders includes biologists, pharmacists, philologists, physicists, engineers, mathematicians, doctors, chemists, architects, sportspeople, musicians, and other artists.[25]

---

19 E. Lamarque, "Alfie Evans Cittadino Italiano. Bene, Certamente. Ma perchè solo lui? [Alfie Evans, Italian. Ok, But Why Only Him?]," *BioLaw Journal – Rivista di Biodiritto* 2 (2018): 43-44.

20 Cittadinanza italiana per meriti speciali a Nosheen Ahmad But [Italian Citizenship to Nosheen Ahmad But because of Special Merits]. http://www.modenanoi.it/cittadinanza-italiana-per-meriti-speciali-a-nosheen-ahmad-but/.

21 https://ricerca.repubblica.it/repubblica/archivio/repubblica/2005/02/07/morto-lazar-berman-grande-interprete-di.html.

22 https://www.firenzepost.it/2014/10/22/firenze-cittadinanza-italiana-al-professor-weiler-presidente-dellistituto-universitario-europeo/.

23 https://www.boe.es/index.php?lang=en introducing the words "concesión de la nacionalidad por carta de naturaleza".

24 J. M. Lete del Río, "Adquisición de la Nacionalidad por Otorgamiento de carta de Naturaleza" [Citizenship Acquired by Extraordinary Naturalisation], *Actualidad Civil* [Civil News] 18 (1996): 407-408.

25 A. Álvarez Rodríguez, "Españoles por carta de Naturaleza: del Privilegio a la Reparación de los Perjuicios Causados" [Spanish by Extraordinary Naturalisa-

However, these are neither necessary nor sufficient conditions, as the appraisal of exceptionality competes only to the Cabinet and is not subject to judicial review.[26] Such estimation should be made upon a case-by-case basis, although there are also naturalisations conferred upon collective recipients.[27]

B. Nature

Ultra-extraordinary naturalisation is often associated to other *ex gratia* powers, such as the presidential pardon or the royal prerogative of mercy.[28] Certainly, both types of measures might be seen as either menaces to the division of powers (lawgivers/politicians and judiciary/government, respectively) or as a "check-and-balance" to remove the flaws of positive law (privilege of naturalisation) or of criminal judiciary (mercy). Whether they should be properly named as rights of grace, discretionary administrative acts, or just simply as political decisions is still a dogmatic cul-de-sac.[29]

In any case, it is undeniable that they both involve a wide discretionary power, admitted by the rule of law as long as it provides some control on its exercise. That has been pointed out, in fact, as the most revealing feature of a system of administrative law.[30]

tion: From the Privilege to the Compensation for Damages], *La Notaría* [Notary] 3 (2012): 42-43.

[26] As settled in a Supreme Court case dated 20th October 2010, which will be commented on ahead.

[27] It is the case of Royal Decree 453/2004, conferring Spanish citizenship to the foreign victims of the 2004 terrorist attacks in Madrid or the Royal Decree 1792/2008, conferring Spanish citizenship to soldiers of the international brigades in the civil war.

[28] G. Cerdeira Bravo de Mansilla, "Nacionalidad por carta de Naturaleza: un Ejemplo de Equidad," 1229-1231.

[29] E. Fliquete Lliso. "Actos Discrecionales, Actos Políticos y actos Graciables: Naturaleza Jurídica del acto de Otorgamiento del Indulto" [Discretionary and Political acts and acts of grace: legal status of the granting of pardon], *Revista Española de la Función Consultiva* 24 (2015): 185-207.

[30] H. W. R. Wade, C. F. Forsyth, *Administrative Law*, 10th edition (Oxford: Oxford University Press, 2009), 286.

Such a (rough) control covers the rules of reason and proportionality,[31] and implies a judicial review only regarding the regulated items behind the political decision, such as procedure or motivation.[32] All in order to avoid a misuse of the act of grace that may lead to arbitrariness.[33] Certainly, as explained by Galligan, the exercise of discretion should be influenced and regulated taking into account values such as stability in legal relationships, rationality in decision-making, due process and fair procedures, as well as other moral and political principles.[34]

From that same point of view, the Spanish Supreme Court has declared, for instance, that judicial review is applicable to a presumed rejection of citizenship conferral based on the privilege.[35] But the devil is in the details and judicial self-restraint regarding this kind of authority is not a surprise.[36] Actually, formal aspects can be controlled, but there is in the end an inalienable discretionary power not allowing a substitution in the assessment of the underlying reasons.[37] Theoretically, once

---

[31] Ibidem, 292-293.

[32] T. R. Fernández, *De la Arbitrariedad de la Administración* [Arbitrariness and Public Administration], 3rd edition (Madrid: Civitas, 1999), 81-96.

[33] Exposing the difference, ibidem, 141-194.

[34] D. J. Galligan, *Discretionary Powers. A Legal Study of Official Discretion* (Oxford: Clarendon Press, 1986), 90-95.

[35] Indictment of 20th November 2008. Contentious-Administrative Chamber. http://www.poderjudicial.es/search/contenidos.action?action=contentpdf&d atabase-
match=TS&reference=3448356&links=%22carta%20de%20naturaleza%22%20Y%20%22nacionalidad%22&optimize=20081230&publicinterface=true

[36] A fact related to the lack of an adequate relationship between control standards and behaviour guidelines, which is essential when discussing the judiciary's role on administrative discretionary decisions as exposed by M. Bacigalupo, *La Discrecionalidad Administrative* [Administrative Discretion] (Madrid: Marcial Pons, 1997), 61-84 or J. M. Rodríguez de Santiago. *Metodología de Derecho Administrativo. Reglas de Racionalidad para la Adopción y el Control de la Decision Administrativa* [Administrative Law Methodology: Rationality Rules for Adoption and Control of Administrative Decision] (Madrid: Marcial Pons, 2016), 24-25.

[37] Judgment of 20th October, 2010. Contentious-Administrative Chamber. http://www.poderjudicial.es/search/contenidos.action?action=contentpdf&d atabase-

the citizenship has been conferred, repeal and *ex officio* review are at the discretion of the state.[38]

The situation is completely different in the case of *subreptio* or *obreptio* because a citizenship inscription based on false or incomplete data may be directly contested before the Civil Chambers upon a more objective basis.[39] Thus, although unusual, appealing against a concession seems more feasible than challenging a refusal (given by an express and motivated resolution or presumed by administrative silence).

Finally, titles to initiate and stand in the lawsuit or the need of an external display of the motives that support the final decision have also been discussed by Italian scholars[40] and before Spanish courts.[41] These arguments have appeared in particular regarding criminal pardon, although in my own view, they could also be extrapolated to citizenship conferred by a discretionary pronouncement. However, despite the Royal Decrees conferring Spanish citizenship under the privilege are officially published, as commented, the text rarely accounts a motivation, which has to be (as it happens in Italy) inferred mainly through press releases.[42] The same happens when referring to the prerogative of mercy.[43]

---

match=TS&reference=5763861&links=%22carta%20de%20naturaleza%22%20Y%20%22nacionalidad%22&optimize=20101104&publicinterface=true.

[38] A. Hernández de la Torre Navarro, "La vía Contencioso-administrativa en la Concesión o Denegación de la Nacionalidad por Residencia. Con Referencia a la Adquisición por Carta de Naturaleza" [Judicial Review of Residence-based Citizenship Conferral or Denial. Special Reference to Extraordinary Naturalisation], in *Problemas actuales del Registro Civil* [Civil Registry Today Problems], ed. José María Bento Company (Madrid: Consejo General del Poder Jucidial, 2007), 209.

[39] J. M. Lete del Río, "Adquisición de la Nacionalidad por Otorgamiento de Carta de Naturaleza" [Citizenship Acquired by Extraordinary Naturalisation], *Actualidad Civil* 18 (1996): 412.

[40] G. Salvini, "La Grazia: Attualità Dell'istituto e Problemi di Legittimità Costituzionale" [Grace: Current Interest and Constitutional Legitimacy], *Rivista italiana di diritto e procedura penale* 3 (1981): 1007-1013.

[41] As accounted and commented in T-R. Fernández, "Sobre el Control Jurisdiccional de los Decretos de Indulto" [Regarding Pardon Decrees Judicial Review], *Revista de Administración Pública* 194 (2014): 209-225.

[42] T. Hualde Manso, "Concesión de la Nacionalidad por carta de Naturaleza. Una Institución y una Práctica Discutibles" [Extraordinary Naturalisation: A

C. Procedure

As previously noted, even if we refer to the privilege of naturalisation as a right of grace, it involves some regulated items suitable for a judicial review, and procedure is one of them. In fact, even if a discretionary decision itself is non-justiciable because it is unsuited for trial-type proof, the administrative decision-making behind it should still be somehow controlled as it must fall within formal standards and be taken according to previously-settled procedures.[44]

Italian regulation is particularly meticulous on this point. The final decision belongs to the President of the Republic and must be contained in a Decree. The initiative must be promoted by the Home Office in consultation with the Minister of Foreign Affairs. Furthermore, the proposal must be discussed by the Council of Ministers and the Council of State must be informed beforehand.

This provision is much more detailed than the Spanish one. There, the Civil Code only takes into account the external shape of the decision (a Royal Decree) and does not need any additional half-way requirement. Complementary regulation disposes that the proposal must come from the Minister of Justice.[45]

D. Effects

Once issued by the President of the Republic, the Decree granting Italian citizenship is presented by the local authorities to the recipient,[46] who within six months has to take a formal oath of loyalty to the Re-

Controversial Institution and Practice], *Revista Aranzadi Civil-Mercantil* 9 (2011): 33-35.

[43] B. Lozano, "El Indulto y la Amnistía ante la Constitución" [Pardon and Amnesty before the Constitution], in *Estudios sobre la Constitución Española. Homenaje al profesor Eduardo García de Enterría* [Studies on the Spanish Constitution. A Tribute to Professor Eduardo García de Enterría], ed. Sebastián Martín-Retortillo (Madrid; Tecnos, 1991), 1049-1051.

[44] D. J. Galligan, *Discretionary Powers. A Legal Study of Official Discretion*, 240-251.

[45] Article 223 of the Decree regulating Civil Register (14 November 1958).

[46] If the new citizen resides abroad, the Decree will be presented by the Consular Authority.

public, and of observation towards the Constitution and the law. Otherwise, the conferral of the citizenship will not become effective.[47]

Spanish regulations operate along the same lines; they even foresee the same due date for the oath (180 days[48]). Spain additionally regulates that only people over 14 years of age must take the formal oath.[49] Such a precaution does not appear in Italian regulations, which only mention that minors (under 18) acquire the citizenship of the parent with whom they live,[50] as if a discretionary conferral directly to them was not envisageable in any case.

III. Discussion on Alfie Evans

The discussion above has formally dissected the regulation supporting the conferral of Italian citizenship to Alfie, providing several comparative law contributions from the Spanish system, which is very similar to the Italian one and has been more often studied in the literature. The rest of the chapter will analyse whether actual facts are in conformity with the law. Such an analysis may bring out new arguments that would enable a stance to be taken on what happened from an unbiased point of view, beyond the activism which tends to coat debates involving the autonomy of the individual, ethics, or religion.

From a public law point of view, the only possible conclusion is that the citizenship award, if ever it existed, was far from being suitable.

---

[47] As disposed by Article 10 of Act 91/1992, which applies to different ways of citizenship acquisition.

[48] Article 21.4 of the Civil Code.

[49] Article 23 of the Civil Code, which has been interpreted *sensu contrario* asserting that the oath is not required for minors under 14. See J. M. Lete del Río. "Adquisición de la Nacionalidad por Otorgamiento de Carta de Naturaleza" [Citizenship Acquired by Extraordinary Naturalisation], *Actualidad Civil* 18 (1996): 409.

[50] Article 14, Act 91/1992. Italian discussions regarding the citizenship of minors focus mainly on second-generation children (born in Italy from alien parents who do not have the automatic right to Italian citizenship). Arguments for and against conferring them citizenship are explained from different perspectives by M. C. Locchi, "Lo *ius Soli* nel Dibattito Pubblico Italiano" [Right of the Soil in Italian Public Debates], *Quaderni costituzionali* 2 (2014): 483-503. The last reform on the matter has been introduced in September 2018, by the so-called Salvini Decree.

And all of that due to reasons which entail formal, material, and structural questions as well as constitutional, administrative and international inflections.

A. The actual existence of the citizenship award and some pro-forma issues

Before going on, a prior facet must be considered. Indeed, a consistent exposition requires first replying to a preliminary and basic question: Did the citizenship conferral really exist? After all the media news, one would be tempted to presume that it was indeed granted, but there are no reports of a formal document supporting the citizenship.

The first news regarding Italian citizenship conferral to Alfie Evans came out on April 23 2018, after a press release by the Minister of Foreign Affairs.[51] The accuracy of such a statement is still weak as it gives account of the conferral by two Ministers (that of Foreign Affairs and the Home Office).[52] It deals with the information as a fact, while, as it was formerly commented upon, the legal basis only provides them with the power to promote an initiative. Obviously, the success is taken for granted, which is understandable to a certain extent because ultimately the decision involved can be seen formally as a presidential one, but it is substantially a governmental act.[53] In any case, this kind of institutional rashness is reprehensible: *Mulier Caesaris non fit suspecta etiam suspicione vacare debet.*[54]

Following the events, the only documentation whose full content is available is an official release of the meeting number 80 of the Council

---

51 *Cittadinanza Italiana ad Alfie* [Italian Citizenship to Alfie] https://www.esteri.it/mae/it/sala_stampa/archivionotizie/comunicati/2018/04/cittadinanza-italiana-ad-alfie.html.

52 In original: "I ministri Alfano e Minniti hanno concesso la cittadinanza italiana al piccolo Alfie. In tale modo il governo italiano auspica che l'essere cittadino italiano permetta, al bambino, l'immediato trasferimento in Italia."

53 R. Clerici, *La Cittadinanza Nell'ordinamento Giuridico Italiano* [Citizenship in Italian Law] (Padua: Cedam, 1993), 330-331.

54 This Latin maxim was coined by Greek historian Plutarco (born at the end of first century) and attributed to Emperor Julius Caesar. In English, it means "Caesar's wife must not only be honest, but also seem so." It is used in the text to express that just the fact of having been under suspicion is enough to not consider somebody (or something) appropriate for the duty of State.

of Ministers, which took place on April 24 2018.[55] According to it, and after a four minute meeting, the Cabinet deliberated the proposal of the Home Office as being in favour of granting Italian citizenship to Alfie Evans.[56] Nonetheless, the final decision must be a Presidential Decree, and it is not at the single disposal of the governmental Cabinet. The press release was again imprecise when identifying the promoters and, on top of everything, there was no trace of the mandatory opinion of the Council of State.

For its part, the official website of Quirinale[57] accounts among the provisions signed by the President during the week of April 23-29 a non-normative Act entitled in a quite coherent way.[58] However, the text of the decision is not accessible, so it is impossible to see whether the legal requirements were fulfilled. The only known data is that it was recorded on the same day (April 24, 2018), increasing suspicions in terms of legitimacy.[59] However, even if the decision ever existed, the citizenship was never effectively acquired, as the conferral was not officially transmitted and formally accepted.

---

[55] *Cittadinanza Italiana ad Alfie* [Italian Citizenship to Alfie]. https://www.esteri.it/mae/it/sala_stampa/archivionotizie/comunicati/2018/04/cittadinanza-italiana-ad-alfie.html.

[56] In original: "Il Consiglio dei ministri, su proposta del Ministro dell'interno Marco Minniti, ai sensi dell'art. 9, comma 2, della legge 5 febbraio 1992, n. 91, ha deliberato il conferimento della cittadinanza italiana ad Alfie EVANS (…)."

[57] Quirinale is the name of one of the famous seven hills in Rome. On top of it is placed the Quirinale Palace, official residence of the President. The name is also given to the President's official website.

[58] Conferimento della Cittadinanza Italiana al Minore Alfie EVANS, nato a Liverpool (Gran Bretagna) il 9 maggio 2016, ai sensi dell'art. 9, comma 2, della legge 5 febbraio 1992, n. 91 [Conferral of Italian Citizenship to the Child Alfie EVANS, born in Liverpool (United Kingdom) on 9 May 2016, based on article no. 9, par. 2, Act no. 91 of 5 February 1992].
https://www.quirinale.it/page/2018_m04d23.

[59] Certainly, any judgment regarding this singular procedure has to be roughly estimated, as exposed by B. Pezzini. "Biopolitica e Biodiritto Nell'atto Singolare di Concessione Straordinaria Della Cittadinanza ad Alfie Evans" [Biopolitics and BioLaw Regarding the Extraordinary Naturalisation of Alfie Evans], *BioLaw Journal – Rivista di Biodiritto* 2 (2018): 55-56.

Ending the formal discussion, the last focal point regards time-opportunity for the use of the privilege and its consistency to the constitutional praxis, as argued by Antonio Ianni: Italy underwent political elections in March 2018 and after the first setting-up of the new Parliament, the government resigned and became interim acting only for *ordinary matters*.[60] Driving this whole extraordinary procedure seems to exceed what it usually considered "ordinary administration" and so it turns out to be contradictory and disharmonic in light of constitutional law.[61]

B. Substantive issues

The previous sub-section tried to explain a set of doubts about the very existence of the citizenship conferral and a review regarding procedure, terms and formal opportunity of the decision. In order to clarify all the points, I presented them as consecutive reproofs, each operating in case the former was not to be accepted. Expanding on the hypothesis, it should be highlighted that there were also material reasons to denounce the citizenship conferral. Furthermore, even if the citizenship conferral had been effective, it would not have been able to produce the expected effects in Alfie Evans' case.

According to the press release of April 23rd by the Minister of Foreign Affairs, the Italian government hoped that the citizenship conferral will enable the immediate transfer of the child to Italy. Nevertheless, that did not make any sense for several reasons.

To begin with, accessing healthcare in an Italian hospital is not limited to Italian citizens. Thus, having a second European citizenship would not change things (as in fact it did not in Alfie's case either).[62] Alfie, as a British child, was a European citizen already entitled with the freedom of movement had it not been for the judiciary prescription firstly delivered by Mr Justice Hayden and then upheld by successive higher English and European courts. The barrier to the transfer was

60 http://www.quirinale.it/elementi/3032.

61 A. Ianni, "La Concessione Della Cittadinanza Italiana ad Alfie Evans: 'Best Interests of the Child' or 'Best Interests of the Government'?" [Italian Citizenship Conferral to Alfie Evans: Best Interests of the Child or Best Interests of the Government?], *BioLaw Journal – Rivista di Biodiritto* 2 (2018): 37-38.

62 E. Lamarque, "Alfie Evans Cittadino Italiano."

never a problem of citizenship but a judicial decision, unassailable at the time, founded on medical reasons and in the best interests of the child.

Furthermore, a change of citizenship does not automatically change the national law regulating minors' status, which as stated in EU rules known as *Brussels IIa* is residence-based.[63] That should have ended any chance of broaching the Italian regulation regarding parental responsibility, which indeed is citizenship-based,[64] but not exempted from litigation.[65] Summing up, parental power is not absolute, and jurisdiction remained in the United Kingdom.[66]

A similar parcours applies to the diplomatic statements indicating the aim to request Evans' immediate transfer to Italy. That may have been useful in case of political or executive obstacles to the transfer of the child, but futile when facing an autonomous judiciary decision by British courts.[67] In fact, in the Vienna Convention on Diplomatic Relations (1930) there is no trace of a right to sue, because it is based instead on the development of friendly relations among nations.[68]

On the other hand, according to the official release of the April 24th meeting of the Council of Ministers, the motivation to confer Italian citizenship to Alfie was the exceptional interest of the *national community* to assure the infant additional therapeutic developments, in the guardianship of relevant humanitarian values that in this case refer to the safeguard of health.[69] It seems then that the Cabinet simply associated

---

[63] See article 8 of Council Regulation (EC) No 2201/2003 of 27 November 2003 concerning jurisdiction and the recognition and enforcement of judgments in matrimonial matters and the matters of parental responsibility.

[64] Article 36 of Act 218/1995.

[65] A. Ianni, "La Concessione della Cittadinanza Italiana ad Alfie Evans".

[66] F. d'Agostino, "Gli sconfortati dibattiti sul *caso Alfie*," 25.

[67] B. Pezzini, "Biopolitica e Biodiritto nell'atto Singolare di Concessione Straordinaria della Cittadinanza ad Alfie Evans," 56-57.

[68] Otherwise there would be support for a *flight of international law* as denounced by J. Ballina Díaz, "Extradición y Funciones del Embajador: el Tribunal Supremo Olvida lo Principal" [Extradition and Ambassador Functions: The Supreme Court Forgets About What is Vital], *Revista Española de Derecho Internacional* 71:1 (2019): 251-256.

[69] In original: "(...) eccezionale interesse per la Comunità nazionale ad assicurare al minore ulteriori sviluppi terapeutici, nella tutela di preminenti

the best interests of the child to the continuation of medical treatment but no evidence was considered or referenced in order to contradict the ones estimated by the British judges.[70]

In addition, the Cabinet identified the interests of the child with the interests of the national community, but the interests of that national community do not need to coincide with those of the state, which is the subject to which the regulation refers.[71] Furthermore, the legal word-for-word reference in Act 91/1992 implies that citizenship conferral responds either to a prize (in case of eminent services to Italy) or to an agreement for mutual benefits (when it resorts an extraordinary interest of the State), but the pretended advantage for Italy is not evident in the case of Alfie Evans.[72]

Finally, the plain motivation is so generic that it may give room to transform a civil regulation into an immigration issue, something that in a decentralised country may risk affecting devolution or administrative responsibilities. Humanitarian values, also mentioned by the Cabinet, raise relevant questions regarding equity. The issue is not trivial as citizenship is normally seen as an instrument for exclusion instead of inclusion.

C. *Structural and collateral concerns*

As already exposed, any decision on the conferral of citizenship has a discretionary scope, exacerbated in the case of extraordinary political decisions regarding special merits or interests. Accordingly, it has been said that while facing sovereign power, there is a desire to obtain citizenship that may be considered as a legally enforceable expectation, but never as a real civil right.[73]

valori umanitari che, nel caso di specie, attengono alla salvaguardia della salute."

[70] J. Harris, "Guarding Charlie and Alfie," *BioLaw Journal – Rivista di Biodiritto* 2 (2018): 8.

[71] A. Ianni, "La concessione della cittadinanza italiana ad Alfie Evans," 39-40.

[72] E. Lamarque, "Alfie Evans Cittadino italiano."

[73] D. Porena, "L'acquisto dello *Status Civitatis* nella Legislazione Italiana: Natura della Fattispecie Provvedimentale e delle Correlate Posizioni Giuridiche Soggettive" [Acquisiton of Citizen Status in Italian Law: The Nature of the Decision and Subjective Legal Situations], in *Cittadinanze*

Therefore, it can be argued that the interest of naturalisation is always that of the state, only subdued to an extrinsic and formal judicial review.[74] This is the reason why the judiciary tends to categorise these kinds of decisions as acts of "High Administration."[75]

Consequently, judicial decisions in terms of citizenship conferral would not pretend anymore to be built on compliance with law, but on consistency with general principles.[76] Hence, the doctrine demanded a more sustainable concept of citizenship, inspired in general principles' rollout.[77] Among those principles, equity is to develop a leading role, as well as the defence of the best interests of the child.

In the context described, the position of the minors is especially weak, something that evokes the debates about the extension of their fundamental rights,[78] especially in case of a collision of their best interests with that of the state. In addition, and in spite of not being the particular aim set in this paper, I must scantly point out that the same arguments may also be exposed and developed in terms of disability. It is in fact a hot topic around citizenship nowadays in Italy, as the Con-

---

Amministrative [Administrative Citizenships], ed. Antonio Bartolini and Alessandra Pioggia (Florence: Firenze University Press, 2016), 362.

74 C. Di Maio, "La Discrezionalità Amministrativa nel "Riconoscere" la Cittadinanza" [Administrative Discretion when Recognising Citizenship], *Diritti Comparati* [Comparative Laws] (2013). http://www.diritticomparati.it/la-discrezionalita-amministrativa-nel-riconoscere-la-cittadinanza/.

75 F. Oliveri, "Cittadini Mancati. La Concessione della Cittadinanza Italiana come Dispositivo Gobernamentale" [Failed Citizens. Italian Conferral of Citizenship as a Government Device], *Studi Sulla Questione Criminale* 1-2 (2014): 104.

76 See, referring generally to discretionary powers, A. Azzena, "Elogio della Discrezionalità Amministrativa" [Praise to Administrative Discretion], in *Los Retos Actuales del Derecho Administrativo en el Estado Autonómico* [Today's Challenges Regarding Regional State for Administrative Law], ed. Luis Míguez Macho and Marcos Almeida Cerreda (Santiago de Compostela: Andavira, 2017), 58.

77 P. Lombardi, "Giudice Amministrativo e Cittadinanza: Quale Contributo per un Concetto Giurdicamente Sostenibile?" [Administrative Judge and Citizenship: How to Build a Sustainable Juridical Concept], *Federalismi.it* 20 (2018): 13.

78 Benito Aláez Corral, *Minoría de edad y Derechos Fundamentales* [Minors and Fundamental Rights] (Madrid: Tecnos, 2003), 87-170.

stitutional Court has recently declared that the lack of a formal exemption to pledge allegiance to the Republic by disabled people is constitutionally unacceptable.[79] Certainly, an absolute obligation to pronounce the oath under a penalty of not acquiring citizenship can be a source of social marginalisation.[80]

To conclude, equity is also an issue to take into consideration, as the whole bureaucratic device needlessly mobilised in the case of Alfie Evans was instead (and still is) not able to react to the situation of many other foreign children for whom Italian citizenship would really make a difference.[81]

IV. Conclusion

Summing up, the Italian move to "save" Alfie Evans can be judged as a shoddy piece of work played to the gallery, rhetorical and useless. The focus of such an opinion, in any case, is not only addressed to Italy's governmental position but also to the social networks and mass media that magnified the political announcements almost as far as producing fake news. The very existence of the citizenship conferral is uncertain.

The Italian public position of some politicians and an official press release risked a diplomatic crisis for nothing, not only due to the final

---

[79] Jugdment no. 258 of December 7, 2017. https://www.cortecostituzionale.it/actionSchedaPronuncia.do?anno=2017&numero=258.

[80] See, commenting the effects of the principle of equality in the legal regime of citizenship conferral and the cited pronouncement of Italian Constitutional Court: S. Rossi. "Incapacitazione e Acquisto della Cittadinanza. Nota a Prima Lettura a Corte cost. n. 258/2017" [Incapacitation and Citizenship Conferral. First Keys on the Constitutional Court's decision no. 258/2017], *Forum di Quaderni Costituzionali* 12 (2017); C. Domenicali, "La "doppia inclusione" dello Straniero Disabile (a margine di Corte cost. n. 258 del 2017) [Double Inclusion of a Handicapped Alien (regarding the Constitutional Court's decision no. 258/2017)]," *Forum di Quaderni Costituzionali* 3 (2018) and P. Addis. "Disabilità e Giuramento per l'acquisizione della Cittadinanza" (Osservazioni a Corte cost., sent. 258/2017) [Disabled People and the Oath to Acquire Citizenship], *Consulta Online* II (2018): 435-446.

[81] L. d'Avack, "Alfie: Quando l'accanimento Terapéutico Diventa una "Pretesa" di Cura" [Alfie: When Therapeutic Persistence Becomes a "Claim" for Care], *BioLaw Journal – Rivista di Biodiritto* 2 (2018): 27.

and unfortunate succession of events which took place on April 28th, but also because, as *supra* exposed, even while Alfie was alive, a citizenship conferral would not have genuinely produced its effects.

Firstly, in order to be effective, a Presidential Decree needed to be officially presented and a formal oath be pronounced. Secondly, had Alfie been Italian, the personal law and jurisdiction applying to his case would not have changed. Finally, the United Kingdom might not have recognised the conferral, arguing that Italy was interfering in a domestic affair due to the lack of a genuine link with Alfie Evans.

This chapter has also evinced the procedural weakness of the administrative action according to Italian national regulation and its inopportune moment from a constitutional law point of view. The material contradictions behind the presumed motives have also been identified and that is a major concern. Certainly, before discretionary powers, the disclosure in advance of standards of decision-making and the disclosure of reasons after the decision has been made are essential in order to repress arbitrariness.

Thus, all these grounds would have been enough to substantiate a judicial review, useful both in a particular and in a general perspective.

Generally speaking, a judicial review would have provided a priceless precedent in a field that demands precision but moves in the utmost uncertainty, fostered by the abuse of undefined legal concepts. In the specific case, it would have clarified at least one key point: the actual existence of the decision awarding Alfie a new citizenship, and its content and motivation.

Chapter 7

Contested Paediatric Palliative Care: A Church of England Perspective

Brendan McCarthy
Church of England, UK

I. Introduction

In its submissions to various government and other consultations, the Church of England has sought to apply a consistent methodology when addressing issues involving medical ethics.

Topics are considered under the headings of safety, efficacy, ethics, and prudence with ethical discussion focused on four cascading moral principles: affirmation of life, care for the vulnerable, creation of a caring and cohesive society, and respect for individuals. Affirmation of life is further understood to reflect not only instrumental (quality of life) and individualistic (personal autonomy) concerns, but also inclusive (relational and social) as well as intrinsic (value) issues.

It is important that contested paediatric palliative care cases are set within this framework for consideration. While every individual case is personally unique, the ethical principles underlying all such cases are similar and share common features with other areas of healthcare.

As part of its commitment to the principles of creating a caring and cohesive society and respecting individuals, the Church of England does not usually make public statements on individual cases where proposed medical care of children is contested by parents. For the same reason, this chapter will focus on principles and practices rather than on specific examples.

II. Background

In contributing to the General Medical Council's consultation on treatment at the end of life,[1] the Church of England supplied a detailed response to the issue of withdrawal of life-sustaining treatment for newborn infants. This response[2] forms the basis of the Church's current approach to contested paediatric palliative care cases and requires to be set out in full:

> Christians believe that all human beings are made "in the image of God."[3] The image of God is not something that a person possesses in the same way as he or she possesses an organ, such as a brain, or faculties, such as the ability to think. The image of God refers to what human beings are:[4] creatures that reflect their maker's character, dignity and purposes. The image of God may be marred, but it can never be erased; part of the purpose of being alive is to experience that image being restored, transforming individuals and communities. It gives humans their distinctive dignity and value; consequently human beings are never to be treated as possessions or commodities.
>
> Christians also believe that human beings, made in the image of God, are made to live relationally, in families, friendships and societies.[5] Communal as well as individual perspectives ought to be considered in ethical decision-making. The "common good" is to be placed alongside individual aspiration. Often, these complement one another, but there are times when individual preferences have to take second place to communal interests just as there are times when individual freedom ought to be upheld over against societal pressure.

---

[1] General Medical Council, *Treatment and Care towards the End of Life* (London: GMC, 2010).

[2] Church of England, "Withdrawal of Life-Sustaining Treatment for Newborn Infants" (2010).

[3] Genesis 1:27.

[4] K. Barth, *Church Dogmatics III/2* (Edinburgh: T & T Clark, 1956), 60.

[5] S. Grenz, *The Social God and the Relational Self* (Louisville, KY: Westminster John Knox Press, 2001).

From these core beliefs emerge principles that guide Christians in making ethical healthcare decisions. These include affirmation of life, care for the vulnerable, the creation of a cohesive and caring society and respect for the individual.[6] Newborn infants are to be accorded the same respect and rights as all other human persons and while there are features of healthcare that are peculiar to newborns, the principles that govern their care are the same as the principles that govern the care of all human persons. There ought always to be a presumption in favour of life and since all newborn infants are vulnerable, they ought to be cared for with respect and dignity. A compassionate society will seek to ensure that all relevant healthcare decisions are made in the best interests of infants, not in the interests of others or because of resource pressures. Consequently, there ought to be an expectation that newborn infants will receive life-sustaining care and treatment that will be continued for as long as it is in their best interests. It is still necessary, however, to examine whether the continuation of certain life-sustaining treatments might not always be in an infant's best interests, whether ethical distinctions may be made between types of life-sustaining treatment and whether healthcare professionals are always well placed to decide what represents optimal neonatal care. Discussions on neonatal end of life care often identify three areas in which it may not be in an infant's best interests for life-sustaining treatment to continue: when treatment is futile, when it is unacceptably burdensome, and when an infant's anticipated quality of life does not justify ongoing treatment.[7]

Genuinely futile treatment is treatment that will fail to alleviate or to cure a particular condition, or treatment that

---

[6] Mission and Public Affairs Division, Church of England Briefing Paper, *Medical Ethics: A Christian Perspective*, 2010, http://cofeweb.zebdev.com/our-views/medical-ethics-and-health-social-care-policy/medical-ethics-.aspx.

[7] Cf. a discussion on the practical outworking of these considerations: "Decisions to Withdraw or Withhold Neonatal Intensive Care", R. Tubman, Consultant Neonatologist, Belfast Health and Social Care Trust, http://www.qub.ac.uk/methics/Tubman.pdf.

will not affect the progress of an untreatable terminal condition while also failing to lessen overall suffering or distress. Continuing to administer treatment in such limited circumstances is futile and withdrawing treatment does not represent a rejection of Christian principles. A fine balance exists between prolonging life and prolonging the dying process and while there will be a grey area in which it may not be clear where this balance lies, the distinction remains a useful one.[8]

Unacceptably burdensome treatment is treatment that will cause a greater degree of overall discomfort or suffering than it will alleviate. A terminally ill infant for whom drug administration involves painful procedures causing the infant to be in notable distress or which necessitates him or her to be continuously sedated conveys no benefit to the child. Such treatment may be withdrawn without compromising the principles of affirmation of life and caring for the vulnerable. It is important to emphasise, however, that the issue here is the burden placed on the infant, not the burden placed on others or on the healthcare "system." No one ought to have treatment withheld or withdrawn because he or she is "a burden" to others.

"Quality of life" issues are often complex, not least because they involve others deciding what quality of life is deemed to be acceptable for the child in question. It is essential in such situations that it is the infant's quality of life that is pre-eminent, not the impact that the child's life will have on the quality of life of others. One mark of a humane and civilised society is that it cares for its vulnerable members,[9] not getting rid of them if their existence becomes an inconvenience to others. Of course, such an approach also necessitates society "caring for the carers."

---

[8] Ibidem.

[9] Cf. The House of Lords Select Committee on the Assisted Dying for the Terminally Ill Bill: *Joint Submission from the Church of England House of Bishops and the Roman Catholic Bishops' Conference of England and Wales,* (para. 4), 2 September 2004.

In order to safeguard the vulnerable, the threshold for acceptable quality of life needs to be set at a low level. The issue is not what quality of life might others find acceptable for themselves, but what sort of experience of life might a child have if life-sustaining treatment were to be initiated or maintained. Given that all children are made in God's image and, consequently, the lives of all children are to be affirmed and protected, only in extreme cases ought treatment to be withdrawn because of anticipated "poor quality" of life. Only where a child has a chronic condition that causes significant, identifiable suffering that cannot be managed apart from an unacceptable level of sedation or where a child is incapable of responding to stimuli in a way that indicates enjoyment, ought the withdrawal of life-sustaining treatment to be considered. To go beyond this is the antithesis of affirming life and caring for the vulnerable. It is society's responsibility to provide proper love and care for its most vulnerable children, not to penalise them further. Christians were at the forefront of outlawing infanticide in the ancient world;[10] they are unlikely to take a different stance to protecting vulnerable children today.

In withdrawing life-sustaining treatment, a distinction may be made between medical intervention and "assisted-care."[11] Providing a safe and therapeutic environment, adequate nutrition and hydration and human contact are fundamental, caring responses to people in need. Even when technological assistance is required in order to provide such care, these responses ought not to be presented as clinical interventions, countering disease or illness. They address basic human needs and they differ in kind from medical, surgical and technical interventions that are designed to address illness. Consenting adults may choose to reject such

[10] C. Horn, "Children and the Experience of Death in Late Antiquity and the Byzantine World," in *Children and Everyday Life in the Roman and Late Antique World*, eds. Christian Laes and Ville Vuolanto (Oxford: Routledge, 2017), 300-317.

[11] Mission and Public Affairs Council, Church of England, *Response to General Medical Council Consultation on End of Life Care*, May 2009.

> care; there are no valid arguments for withdrawing these caring responses from newborn infants other than where their provision would actively increase a terminally ill child's suffering and distress.

The issue of who decides what is in a child's best interests is a vexed one. Health professionals cannot be forced to continue treatment that they consider to be futile or unacceptably burdensome, but the opinions of parents ought fully to be taken into account. If agreement is not evident and cannot be attained, the decision on how to proceed may pass to the courts. It is often the case that parents wish life-sustaining treatment to be continued in spite of a prognosis of a severely limited quality of life for their child. In such circumstances, parental opinion ought to be given precedence with the onus being on others to seek court intervention.[12]

In summary, the Church of England's principled approach is one that seeks to affirm life, to care for the vulnerable, to build a caring and cohesive society and to respect individuals. Consequently, in practice, life-sustaining treatments ought to be withheld or withdrawn from newborn infants only in very limited circumstances.

III. Contested care

In a relatively small number of cases, healthcare professionals and parents disagree with regard to the best pathway for the ongoing care of seriously, critically or terminally ill children. Given the discussion above, how might the Church of England approach these sad and difficult cases?

A. Safety

The first question to be asked is, which is the safest practicable approach to be taken that is in the best interests of the child? This is not simply a matter of deciding which interventions will prolong life or postpone death. Safety includes minimising suffering and distress as well as enhancing comfort and wellbeing.

For example, a proposed course of treatment might offer some prospect of prolonging life or of alleviating some symptoms of illness, but

[12] Ibidem.

it might also carry a risk of having contrary effects. In such cases, clinicians have to balance potential benefits against potential harm. Such a calculation might often be complex to make, but in the real-life world of healthcare it is one that it is necessary to grasp. It is also one that only medical experts can make, based on the best clinical evidence available.

It is the role of medical experts to identify the levels of risk associated with potential interventions and to propose treatment accordingly; an agreed course of action taken might be more difficult to establish as will be discussed later.

B. Efficacy

In a number of cases where health professionals and parents have disagreed on the treatment that a child should be given, novel or experimental treatments some of which have been available only outside the UK, have been favoured by parents.

In such cases, it is the role of medical experts to assess not only the safety of such treatments, but also their efficacy. It is entirely understandable that parents might wish their child to "be given every chance" no matter how unlikely success might be or for them to believe unconfirmed or exaggerated accounts of the efficacy of suggested interventions. Such situations might be exacerbated by claims being made in jurisdictions where governance is less robust than the UK or by "rogue" practitioners. The predicament of parents is made all the more acute by occasional reports of successful treatment against the expectations of medical experts, even though the facts might be more complicated than the headlines suggest.[13]

As with safety, however, it is the preserve of medical experts to come to a decision on the likely efficacy of potential treatments and to base their clinical judgement on that assessment. Again, as in considerations of safety concerns, finding an agreed course of action might not be as straight-forward as coming to a clinical judgement; this too will be discussed in the section on Prudence.

---

[13] J. Adams, "Ashya King Cleared of Cancer Three Years After his Parents Abducted him from Hospital for Treatment Abroad," *The Telegraph*, 3rd March 2018. https://www.telegraph.co.uk/news/2018/03/03/ashya-king-cleared-cancer-three-years-parents-abducted-hospital/ (accessed December 5, 2018).

C. Ethics

The Church of England's four cascading principles of affirming life, caring for the vulnerable, creating a caring and cohesive society and respecting individuals are key to exploring cases of contested care.[14]

Affirming life must begin with a "presumption in favour of life" which is usually understood as saving or prolonging life. This presumption, however, is not absolute and ought to be informed by various ways in which life is understood to be valued.

It is common for the value of a person's life to be understood in terms of "quality of life" which often focuses on either instrumental or individualistic ways of determining how a person's life ought to be viewed. It is, of course, important to consider which functional abilities a child has (or is likely to have in the future), but there is a danger of health professionals and others (including parents) judging quality of life in this way, based on their own functional abilities which are much more fully developed. So, for example, it might be reasonable for an adult to think that their quality of life would be negligible if they were unable to see, hear, talk or walk, but it does not follow that a child who knows no other way of living has a negligible quality of life.

Similarly, while adults might be able to state that it is their own view of the value of their life that matters most and so they ought to be free to make advanced decisions, that avenue is not available to young children; great care ought to be taken in exercising rights on their behalf.

All human beings are part of networks of family and community and the value of an individual life is also to be understood in this context. In cases of contested care, the value and importance of a child to its parents must not be underestimated. Parents are not "being selfish" in wanting their child to live "against the odds;" their desire to see their child live is an expression of the value that child has.

Most importantly, every human being is of intrinsic value, regardless of their abilities, aptitudes or even their self-estimation. This principle forms the basis of human rights law and underpins much of criminal law and could only be set aside at society's peril.

---

[14] B. McCarthy, "Medical Ethics: A Christian Perspective" (3 November 2017). https://www.churchofengland.org/sites/default/files/2017-11/Medical%20Ethics%20a%20Christian%20Perspective.pdf.

These reflections suggest that to affirm life in the context of neonatal care is to have a strong presumption in favour of saving and prolonging life, but other principles also need to be considered.

Caring for the vulnerable might be seen to favour saving or prolonging life in most circumstances. While this is a correct starting point, a second, important factor is the alleviation or avoidance of suffering and this must be given due weight in decision-making.

The degree to which acting to save or prolong life might cause suffering is often difficult to discern. Suffering is a subjective concept as is the experience of pain; what might constitute unbearable pain or suffering for one individual might be well within the limits of acceptability for another. Healthcare professionals can only estimate the degree of pain or suffering that a treatment might cause and have to interpret physical signs as indicators of suffering in children who are unable to communicate.

A more complex issue is "burdensome treatment." This often refers to treatment that will not significantly alter the progress of illness or disease, but which, in itself, might cause discomfort. It is, essentially, treatment that postpones death. To what extent such treatment might be in the best interests of a child is a contested point even though such treatment might be deemed to be helpful to parents who are also vulnerable individuals in these circumstances.

In essence, caring for the vulnerable, in most circumstances, will mean acting to save or prolong life, but in cases of paediatric palliative care, caring for children might mean withholding or withdrawing some treatments in order to avoid or alleviate suffering. Such a decision is difficult to make and is often not as clear-cut as making decisions on the safety or efficacy of proposed treatments.

Creating a caring and cohesive society entails a recognition that individual cases must be set in a wider societal context. This means that the way in which contested cases are dealt with by healthcare professionals and, when necessary, by the courts, must both demonstrate real care to all involved (children, parents and staff) while at the same time ensuring that the overall "message" given to society is one that will contribute to it becoming more, rather than less, cohesive.

This is difficult to achieve as individual cases might sometimes suggest that a particular course of action might be compassionate, but would have a detrimental effect on others. For example, a particularly difficult set of circumstances might see healthcare professionals and even parents wishing that euthanasia was permissible, but wisely, in the UK euthanasia remains unlawful,[15] not least because to permit it in exceptional circumstances would be to breach a principle that once breached could not be closed again.

In brief, how contested cases of paediatric palliative care are treated ought to help create a more compassionate society; if a course of action fails this test, it requires re-examination.

Respecting individuals in contested cases necessitates a covenant of trust to be established between healthcare staff and parents. In some recent high-profile cases, this covenant clearly broke down with social and mainstream media exacerbating rather than helping resolution of disagreement.

Similarly, external organisations and spokespersons sometimes join a very personal debate, often with little or no detailed knowledge of the circumstances. Apart from those directly involved in the care of seriously or terminally ill children, few will be aware of the complexities of their situations. Respecting individuals necessitates people who are external to the case not making public statements or "taking sides" as if all concerned were not trying to do what they consider to be in the best interests of the child - one reason why the Church of England does not normally issue statements on individual cases (and why this chapter will not attempt to comment on them either).

D. Prudence

Even when medical experts have come to a clear opinion with regard to the safety and efficacy of proposed treatments and even when they are assured that, within the ethical guidelines proposed above, it is in the best interests of a child to have life-saving or life-prolonging treatments to be withheld or withdrawn, parents might strongly disagree. While this is a relatively rare occurrence, given the number of tragic cases that arise every year in the UK, how might such situations be dealt with in a compassionate and prudent manner?

It is important to try, first of all, to separate fact from opinion even though this delineation might not always be perfect.

---

[15] Section 2(1) Suicide Act 1961.

Matters of safety and efficacy are topics for expert assessment and must not be allowed to become issues of personal or public opinion. Experts do, of course, differ from time to time, but this is seldom the case in assessing major questions of safety or efficacy in instances of paediatric palliative care. Where genuinely differing expert opinion exists, the difference ought to be dealt with as detailed below.

One vexed issue is that in some cases, parents might appeal to an "expert" who lacks the qualifications and experience to be deemed as such. Expert assessment ought to mean assessment by senior, practising clinicians within the UK; it is essential that current (and hopefully, transient) popular distrust of experts is not allowed to permeate this area if the best interests of children are to be upheld.

If it is accepted that medical experts alone are competent to decide matters of safety and efficacy, how best might those decisions be shared with parents who understandably might wish that they had come to different conclusions?

It is at this point that issues of partnership and communication come to the fore. It is essential that, wherever possible, an "us and them" atmosphere is not allowed to develop within a paediatric ICU. From the outset, healthcare professionals and parents must know that they are on the same side. In order to minimise the possibility of such relational breakdowns, it is imperative that specially trained nurses are employed in all paediatric ICUs to act as liaison professionals between clinicians and parents. Following the model of Specialist Nurses for Organ Donation, such nurses can get alongside parents from the beginning of their child's treatment, explaining and mediating clinical decisions.

In helping parents and clinicians to agree on treatment options, an early informal contract of trust ought to be negotiated where clinicians agree to act in favour of saving and prolonging a child's life except when it is agreed that to do so would be to cause or to prolong pain or suffering that could not, themselves, be alleviated and which would not alter the underlying condition, or where a child has severe irreversible brain damage so extensive that minimal consciousness is, and would remain, absent. In practice, this might mean that some children will be given treatment for a longer period of time than is strictly necessary, but once it is accepted that pain and suffering will be prolonged needlessly by further treatment, palliation alone ought to be pursued.

It must be stressed that palliation does not equate to removal of treatment; it signifies a different type of treatment to that aimed at finding a cure.

Where further treatment will not cause pain or suffering, but will be burdensome in the sense that it will prolong the inevitable dying process, sensitive decision-making is called for. If the level of discomfort caused by continuing treatment is deemed to be small, the benefit to parents ought to be given serious consideration. If discomfort comes close to causing pain or suffering, a strong case for withdrawing or withholding specific treatments can be made. The best interests of the child must always be paramount, but where best interests are not compromised by continuing treatment, the care of vulnerable parents must also be given due weight.

In most cases, a specialist liaison nurse will be able to walk parents through the stages of care with compassion and understanding, enabling them to act as partners with clinicians.

Where this is not possible, some may suggest that external "second opinions" ought to be sought or that parents and clinicians should undergo mediation. The evidence for these strategies producing positive outcomes is limited and can, in fact, prolong disagreements between parents and healthcare professionals. If specialist liaison nurses were to be involved at every stage of a child's care, working alongside other health professionals and parents and disagreement between parents and clinicians still ensued, it is unlikely that mediation or involving a further external expert would lead to a better outcome.

Sadly, in those small number of cases where a relationship between clinicians and parents still breaks down in spite of the best efforts of a liaison nurse, redress to the courts is the last resort.

There is a strong case to be made that in order to respect individuals, including staff as well as parents, that reporting restrictions ought to be put on such cases so that individuals are not named or identified. Equally, it ought to be made clear that judgments are made with regard to a child's best interests; they ought not to suggest a determination that "one side" has won and the other lost.

Equally, the courts ought to be careful in the language they use to communicate their decisions. In particular, they must be clear that a decision to withhold or to withdraw specific treatment is not the same as withholding or withdrawing treatment. Medical and nursing care will always be given to seriously and terminally ill children; the issue is not removal of care, but determination of the best course of care that is in a child's best interests.

IV. Conclusion

There are no easy answers or remedies to be found in cases of contested paediatric palliative care, but emphasis ought to be placed on trying to avoid breakdowns in relationships between healthcare professionals and parents in the first instance.

Following a clear pathway where issues of safety, efficacy, ethics, and prudence are followed in a transparent manner with the involvement of specially trained liaison nurses might not remove the need for involvement by the courts entirely, but it ought to go some way in achieving this goal.

Chapter 8

The Dynamics of Clinical Judgment, Religious Convictions and Parental Responsibilities: An Islamic Perspective

Mahmood Chandia,[1] Abdulla al-Shami[2]

[1] University of Central Lancashire, UK
[2] The Tessellate Institute, Ontario, Canada

I. Introduction

The latter part of the 20th century and the first part of the 21st century have challenged not only the state and health professionals with balancing medical treatment against ethical dilemmas and financial prudence, but also parental (in the case of a minor) or patient autonomy. This is all the more evident in cases where treatment is deemed "medically futile." Moreover, this issue becomes acutely nuanced with sensitivities when the state or medics intervene or override parental authority to discontinue medical treatment. In such a scenario, the state and medical professionals determine the course of treatment even for a minor. This is tantamount to assuming absolute autonomy and rendering parental autonomy ineffective. This latter consequence of state intervention is of particular concern. Since it has implications on how faith-based communities discharge their parental responsibilities and to what extent this is or is not compatible with religious values and instructions. This is a significant matter, given that such values pre-date the advancement of modern technology.

The present chapter is an attempt to appreciate the dynamics, tensions, and challenges of such scenarios, particularly where children are concerned, for at least one Abrahamic community, the Muslim community. Likewise, this chapter aims to examine what directives are offered to balance secular legal and clinical viewpoints against Islamic legal understandings. This issue has gained more prominence in the recent past due to two high profile cases, namely the cases of Charlie

Gard and Alfie Evans. Both have once again raised the debate over the position of parental authority against state and medical authority.

The essential purpose of this chapter is to evaluate the dynamics of negotiating a third space between clinical practice and religious conviction on parental duties from an Islamic perspective. Indeed, modern day technology has revolutionised clinical diagnosis and practice but it has also raised many ethical dilemmas. At the heart of this chapter is a two-fold aim: 1. What is the scope of parental rights and responsibilities in Islam (i.e. theory)? 2. How would Islamic *fiqh* (legal formulation) guide parental thinking in such situations (i.e. discharging parental responsibility which equates to practice)?

II. Parental rights and responsibilities in Islam

Islam from the outset has established a very lofty social status for not only biological parents but also for step-parents. Alongside affording parents a set of rights, it has also subjected them to shoulder responsibilities. The latter is more pronounced during the formative phase of a child's development. The dynamics of such a relationship between parents and children is tantamount to not only a social but also a spiritual covenant. The Qur'ān, the primary scripture of Islam, offers instructive guidance to its readers and followers on the nature of such covenants, namely that each party of a covenant is accountable for how it discharges and executes its duty[1]. In fact, both parties (parents and children) have rights and responsibilities towards and over each other.[2] Without a doubt, being accountable for the discharge of responsibilities is subject to mental sanity and faculties. Both rights and responsibilities are explicated from the epistemological framework of Islam, i.e. its primary scriptures and secondary sources.

To appreciate the spiritual and social status of parents, the Qur'ān establishes an underpinning principle that extending the utmost reverence towards parents in speech and action is godliness. In fact, the Qur'ān, in several places where the association of partners with the One Creator is utterly condemned in the strongest of terms, also juxtaposes this with an injunction to extend excellence in behaviour

[1] Al-Qur'ān, Sūra al-Isrā, verse 34.

[2] Al-Qur'ān, Sūra al-An'ām, verse 151 and Sūra al-'Isrā, verses 23 and 31.

towards parents; irrespective of faith indifferences.[3] For example, at the latter place, the Qur'ān guides:

> And if they impose upon you that you associate as a partner with Me what you have no knowledge of, then do not obey them. And accompany them in the world with kindness, as per customary practice, and follow the way of the one who turns to Me in repentance. Your return is to Me; I will then inform you of what you used to do.

The juxtaposition of this topic alongside the central creed of Islam is ample testimony for the believer to cognise the social status of parents and comprehend their spiritual status. The Qur'ān further records:

> Your Lord has decreed that you only worship Him; and that you do goodness toward parents. If one of them or both of them reach old age with you, then do not say to them, '*Uff*' and do not scold them, rather say a kind word to them.[4]

The Qur'ān has repeatedly emphasised a similar sentiment in different places.[5] Such repetitive emphasis underpins its significance. The Qur'ān also guides that children are a trial which can potentially divert from higher purposes: "Your wealth and your children are only a trial and there is an immense reward by Allāh."[6]

This significance and the social reverence of parents is also a recurring feature of Prophetic narratives, commonly referred to as "Hadith," in Islamic terminology, where this phenomenon is highlighted in variant forms. For example, in one such Hadith, it is explained that compromise on extending reverence to parents has both temporal and eschatological implications. The Prophet of Islam is reported to have said: "Should I not inform you of the most-grave of the major sins?" He then went on to explain three behaviour patterns: (a) the association of partners with the One Creator; (b) to knowingly

---

[3] Al-Qur'ān, Sūra al-'Ankabūt, verse 8 and Sūra Luqmān, verse 15.

[4] Al-Qur'ān, Sūra al-Isrā, verse 23.

[5] Al-Qur'ān, Sūra al-Baqarah, verse 83; Sūra al-Nisā', verse 36; Sūra al-An'ām, verse 151; and Sūra Luqmān verse 14.

[6] Al-Qur'ān, Sūra al-Taghābun, verse 15.

provide a false testimony; and (c) to be undutiful towards parents.[7] In another Prophetic narrative, it is explained that "[t]he best deeds are [the performance of] prayer at its prescribed time and kindness to parents."[8] Likewise, in another Prophetic narrative, it is stated that "[t]he Lord's pleasure is certainly in parental happiness; and the Lord's anger is in parental dissatisfaction."[9]

This topic is also elaborated upon with practical exemplification in Islamic legal discussions. These Islamic legal discussions also encompass human social conduct. Muslim jurists, whose role is to provide legal guidance and ensure living Islam is possible across time and space albeit subject to the values of the Islamic epistemology, have also extensively discussed parental rights and responsibilities. To the extent that practical suggestions are also offered. These include, being dutiful to one's parents,[10] offering financial support,[11] expressing obedience to them,[12] supplicating for them,[13] and greeting their associates with respect,[14] etc.

---

[7] Al-Bukhārī, *Şahīh al-Bukhārī* [The Authentic Collection of Imam Bukhari] Book 87, Hadith 9; Al-'Asqlanī, *Fatḥ al-Bārī* [The Opening from the Creator], p. 565.
http://library.islamweb.net/newlibrary/display_book.php?flag=1&bk_no=52&ID=12229.

[8] Muslim, *Şahīh* Muslim [The Authentic Collection of Muslim] Book 1, Hadith 162.
http://library.islamweb.net/newlibrary/display_book.php?idfrom=274&idto=280&bk_no=53&ID=45.

[9] Al-Mubarak Fūrī, *Tuḥfat al-Ahwadhī* [The Gift of al-Aḥwadhī, Commentary on the Sunan of al-Tirmidhī], vol. 1, 23 (Dār al-Fikr, Beirut, no date of publication).

[10] Al-Bukhārī, *Şahīh al-Bukhārī* [The Authentic Collection of Bukhari], vol. 7, p. 20; Hadith no. 5080, Dār al-Ta'ṣīl [House of Establishing the Origin] (Cairo, 2012).

[11] "Your children are part of the best of your earnings, so eat from what your children earn." *Sunan Al-Nasā'ī* [The Sunan Collection of Imam Nasā'ī,].

[12] Al-Qur'ān, Sūra al-Isrā, verse 23.

[13] "My Lord, have mercy on them just as they compassionately cared for me when little." Al-Qur'ān, Sūra al-Isrā, verse 24.

[14] Muslim, *Şahīh* Muslim [The Authentic Collection of Imam Muslim] Book of Faith. Hadith 2552 (Dār al-Salām, Cairo, No date of publication).

The statements above establish that while respectful behaviour with parents is a highly laudable deed, contra behaviour to that is a grave sin. Ultimately, Divine pleasure can be secured via the former route. Such is the merit of dignity of duty to parents and irreverence to them is not only civil contempt but can attract grave consequences – both temporal and eschatological.

Nevertheless, this instructive guidance is balanced by a sense of responsibility for offspring to be shouldered by parents. This is highlighted in quite a succinct manner in the following Hadith in which parental responsibilities are explained. The Prophet has said, "Just like your father has a right over you; your child has a similar right."[15] The Islamic legal experts further explain this social duty by adding that parents' responsibilities towards their child include: selecting or offering wisdom in selecting a good spouse;[16] choosing a good name for the child; nurturing the child in a good manner and educating him/her; guiding, helping and supporting the child in doing good deed; treating siblings with equality and fairness; etc.[17]

In summary, parental rights also attract parental responsibilities. Whilst offspring are guided to be dutiful of parental rights, parents are equally guided to be conscious of being accountable in discharging responsibilities.

---

http://library.islamweb.net/newlibrary/display_book.php?idfrom=274&idto=280&bk_no=53&ID=45.

[15] Ali Al- Haithami, *Majma' al- Zawā'id* [A Collection of Additional Hadith] vol. 8, 146. Hadith no. 13422 (Dār al-Salām, Cairo, No date of publication).

[16] "A woman is married for four things: for her wealth, for her lineage, for her beauty or for her piety. Select the pious one, may you be blessed!" Al-Bukhārī, Book of Marriage, Hadith. 4725 (Dār al-Salām [House of Peace] Cairo, No date of publication).
https://library.islamweb.net/hadith/display_hbook.php?bk_no=146&pid=103269&hid=4725.

[17] For more reading, see A. Alwāni. *Tarbiyat al-Awlād fī al-Islām* [Nurturing of Children in Islam] (Dār al-Salām, Cairo, 1992).

III. Understanding medical futility and discharging parental duty

A. Understanding medical futility and the variant perspectives

Medical futility is either understood in a subjective manner based on quality of life or in an objective manner based on physiological parameters.[18] The debate revolves around a set of concepts: medical benefit, responsible practice of medicine, duty of care, omission, best interests, patient autonomy, etc. The physician's duty to apply or not medical treatment (i.e. clinical judgment) is balanced against the patient's rights. This can lead to intricate decision-making and ethical dilemmas. Without a doubt, each decision has an economic implication too; whether the cost is borne by private or public resources. There is an imposition on finite state resources and the potential to divert valuable resources away from other patients where the prognosis of treatment and survival is better. This debate has attracted competing viewpoints. Drane and Coulehan argue that physicians, ethicists, the state, and religious communities all have a distinct perspective. All of them contest what is socially sanctioned standard of rationality and what is sacred.[19] A closer and critical evaluation of the variant viewpoints can be an overwhelming or at least a daunting task. But they provide a very informed overview and conclude that "clarifying the concept of futility and establishing defensible ethical policies covering futility are important steps toward eliminating unhelpful, medically inappropriate practices."[20]

B. Understanding the Islamic perspective on medical futility

The concepts of "medical futility" and "death" have attracted contrasting Islamic legal (*fiqh*) opinions within Muslim scholasticism. One of the central issues that divides this range of opinion is the "concept of death" in Islam. So, to grasp what the Islamic perspective

---

[18] M. A. Albar, H. Chamsi-Pasha, "Futility of Medical Treatment," *International Journal of Human and Health Sciences* 2:1 (2018): 14.

[19] J. F. Drane, J. L. Coulehan, "The Concept of Futility. Patients do not have a Right to Demand Medically Useless Treatment. Counterpoint," *Health Prog.* 74:10 (1993): 28-32.

[20] Ibidem.

is on determining what medical futility is, necessitates a critically-informed comprehension of the understanding of death.

Islam defines death by irreversibility. This includes the cessation of all vital biological functions. However, medics define death by a subjective criteria based on neurological disability resulting from severe brain injury such as coma and dependency on mechanical ventilation for breathing (apnea) and the quality of life.[21] This arbitrary definition of death, coma, and apnea enable organ donation and transplantation from dying or terminally ill patients. In Islam, the absence of life defines death.

The religious definition of death which calls for the sanctity of life and the inevitable occurrence of death for all mankind has been clearly stressed in Islam. A study of a few verses provides an insight into this understanding. The Qur'ān states that "Every soul shall taste death …"[22]; further, "You will certainly die and they will certainly die,"[23] hence, the permission for a soul to die is absolutely subject to Divine will in Islam, and "It is not for a soul to die except at a written time with the permission of Allāh[24] and

> Allāh takes the souls away at the time of their death and those who have not died in their sleep. He then keeps back those upon whom He had decreed death and He releases others to a fixed predetermined time. In this are most certainly signs for a people who reflect.[25]

While the inevitability of death and who predetermines it, are made fairly clear in the above verses, the exact definition of death remains vague. Just as the inevitable occurrence of death is well defined, the sanctity of human life has also been abundantly stressed in Islam in several verses.[26] Islam advocates that every effort should be made to

---

[21] F. A. Khan, "The Definition of Death in Islam: Can Brain Death Be Used As A Criteria Of Death In Islam?," *The Journal of IMA* I8 (1986): 18-21 at 20.

[22] Al-Qur'ān, Sūra al-Āli 'Imrān, verse 185.

[23] Al-Qur'ān, Sūra al-Zumar, verse 30.

[24] Al-Qur'ān, Sūra al-Āli 'Imrān, verse 145.

[25] Al-Qur'ān, Sūra al-Zumar, verse 42.

[26] Al-Qur'ān, Sūra al-Baqarah, verse 195; Sūra al-Nisā', verse 29; Sūra al-Mā'idah, verse 32 and Sūra al-'Isrā verse 31.

preserve life, and take good care of the body; death will occur at a predetermined time.[27]

All Islamic legal works define death as the departure of the spirit *(al-rūḥ)* from the body. The Qur'ān discusses the spirit[28] but emphasises that its nature is beyond human understanding.[29] However, the Hadith narratives do provide some clarity in defining death. This is based on two complementary Hadith: (a) "When the soul is taken away, the human sight follows it"[30] and (b) "When you attend to your dead ones, close their eyes, as the sight follows the soul. And say good things, as the Angels say 'Amen' to what the members of the household say."[31] Both Hadith underscore that the notion of death is to be understood as being in a lifeless state as the soul is departing the body.

Jurists have further explicated within exegetical and Hadith-related literature that the following can be taken as determinant signs of death: when the human body stops breathing, when the feet loosen, palms hang loose, the nose drops to one side, … and the body becomes cold.[32]

---

[27] F. A. Khan, "The Definition of Death in Islam," 19.

[28] Al-Qur'ān, Sūra al-Isrā, verse 85.

[29] Ibidem. Commentators understand *al-Rūḥ*, the spirit, in various ways. It seems that in this context, it refers to the human soul. This verse is said to have been revealed in answer to the third of three questions which some Jewish rabbis prompted the pagans to ask. The other two questions are answered in the introduction to Sūra al-Kahf. See D. W. Peachy, M. H. Al-Johani, *The Qur'an: The Final Book of God, A Clear English Translation of the Glorious Qur'an*, Qassim, 177 (World Assembly of Muslim Youth, Riyadh, Kingdom of Saudi Arabia, 2010). This book is the reference of the English translation for any verse cited in this research.

[30] Muslim, *Ṣaḥīḥ Muslim* [The Authentic Collection of Imam Muslim], vol. 2, 635, hadith no. 1529 (Dār al-Salām [House of Peace], Cairo, no date of publication).
http://www.islamweb.net/maktaba/index.php?flag=1&page=bookpages&bookid=1&id=1601&bookparts=[635:2]&LoadTab=LoadBookDetail.

[31] Ibn Mājah, *Sunan Ibn Mājah* [The Sunan Collection of Ibn Majah], vol. 1, p. 467, hadith no. 1522 (Dār Iḥyā' al-Turāth, Cairo, no date of publication).

[32] Ibn 'Ābidīn, *Rad al-Muḥtār 'alā al-Dur al-Mukhtār* [The Preferred Response to the Preferred Gem] known as *Ḥāshiyat Ibn 'Ābdīn* [The Gloss on Ibn Ābdīn], vol. 1, p. 89, 'Ālam al-Kutub [The World of Books] (Riyadh, Kingdom of Saudi Arabia, 2003); *al-Fatāwā al-Hindiyah* [Legal Edicts of al-Hind i.e. pre-partition

Abū Ḥāmid al-Ghazalī (d. 1111) stated that:

> The meaning of the departure of the spirit from the body refers to when the body does not respond to any command given by the spirit, since all bodily parts are simply tools for its use. Therefore, death occurs at a time when parts of the body cannot respond.[33]

He further pointed that the Qur'ān clearly states that death does not happen except by Divine decree. As mentioned in the Qur'ān: "Blessed is He in Whose hand is power and He is Omnipotent over everything. He, Who has created death and life in order to test who from among you is the best in deeds."[34]

It is understood, in light of this verse, that life is a Divine trust. Only He can determine its existence or otherwise in a person. Given this, some scholars actively do not support medical treatment if the purpose is to simply prolong the final stages of life. Life is a Divine trust and cannot be terminated, they contend, by any form of active or passive human intervention, as its term is absolutely determined by an inalterable Divine decree.[35]

India] vol. 1, p. 154 (Dār Iḥyā' al-Turāth [House of Resurrection of Heritage] Cairo, no date of publication); Khalīl, *Mukhtaṣar Khalīl* [The Epitome of Khalīl] vol. 1, p. 37 (Dār al-Fikr [The House of Ideology] Cairo, 1981); M. al-Nawawī, *Rawḍa al-Ṭālbīn* [The Garden for Students] vol. 2, p. 98 ('Ālam al-Kutub [The World of Books] Riyadh, Kingdom of Saudi Arabia, 2003); M. al-Sharbīnī, *Sharḥ al-Minhāj* [The Commentary on the Path] vol. 1, p. 322 (Dār al-Fikr [The House of Ideology] Cairo, no date of publication); Ibn Qudāmah, *al-Mughnī*, vol. 2, p. 452 ('Ālam al-Kutub, Riyadh, Kingdom of Saudi Arabia, 1997); M. A. Ibn al-Najār, *Muntaha al-Irādāt* [The Final Destination of Intentions], vol. 1, p. 323 ('Ālam al-Kutub [The World of Books] Riyadh, Kingdom of Saudi Arabia, 1999).

33 M. M. Abū Ḥāmid al-Ghazalī, *'Iḥyā' 'Ulūm al-Dīn* [The Resurrection of Religious Sciences], vol. 5, p. 166 (Dār al-Ḥadīth [The House of Hadith] Cairo, 2004).

34 Al-Qur'ān, Chapter 67, verses 1-2. See also K. A. Choong, M. Chandia, "Technology at the End of Life: 'Medical Futility' and the Muslim PVS Patient," *International Review of Law* 9 (2013), available at http://dx.doi.org/10.5339/irl.2013.9.

35 M. A. Albar, H. Chamsi-Pasha, "Futility of Medical Treatment," *International Journal of Human and Health Sciences* 2:1 (2018): 16.

To this end, based on the Prophetic Hadith which states that "[t]here should not be causing of harm nor reciprocating harm,"[36] Islamic jurists debate that any omission of medical treatment toward any patient should be conducted after a robust consultation with the patient and his family, and any other interested third party in the welfare of the patient.[37] The following legal maxims, they argue, can also be applied to determine a course of action: "Necessity can permit prohibitions," "Hardship begets ease," "Doubt does not refute certainty." The application of these maxims alongside others can assist in negotiating a third space, i.e. a common ground for action. However, they argue that the withdrawal of treatment does not mark the point at which any fixed portion of inheritance or the will for the Islamically legal distribution of the deceased's estate can be executed until the spirit has completely departed the body. In short, "omission" can be considered but this is not to be taken as death.

Furthermore, this should not compromise in any way, within the Muslim patient or Muslim relatives, the integral belief that the absolute source of all healing is ultimately God, and that any cure is always subject to Divine will. Nevertheless, it is equally important to believe that, if God wills, cure can be achieved but supplicating for a miracle does not obligate Muslims to demand treatment if experts have deemed it to be futile. Consequently, Muslims can allow for life to take its natural cycle, as death is considered a passage to eternal life and meeting one's Creator. However, a religious expert should ideally be consulted to address any religious concerns the family may have.[38] It should also be noted that withdrawal of any life-sustaining treatment, when considered futile, would be seen as "allowing death to take its natural course; delaying the inevitable…is neither in the patient's nor in the public's best interests because of distribution of healthcare and financial resource."[39] According to Islam, the physician needs to be

[36] Ibn Mājah, *Sunan Ibn Mājah* [The Sunan Collection of Ibn Mājah] 2, 63, hadith no. 2340 (Beirut Dār Iḥyā' al-Turāth [House of Resurrection of Heritage] Cairo, no date of publication), http://library.islamweb.net/newlibrary/display_book.php?bk_no=54&ID=795&idfrom=4293&idto=4422&bookid=54&startno=32.

[37] M. A. Albar, H. Chamsi-Pasha, "Futility of Medical Treatment."

[38] Ibidem.

[39] Ibidem.

certain of the inevitability of the impending death otherwise life should be sustained.[40] The functioning of the body is a strong determinant in this decision-making process.[41]

In essence, the Islamic definition of death is the departure of the spirit from the body, although the science of medicine defines it as the loss of "vital bodily function."[42] It seems they are not necessarily the same, but there is a degree of commonality that facilitates constructive grounds for discussion and action.

C. Understanding Muslim parental responsibility for terminally ill patients

It is always a challenging time when parents are relayed a poor health recovery prognosis for their child from ill health. At such times, where the illness is considered terminal or life-threatening, or one that may well severely impact the quality of life, parents of children from faith communities more often than not, very conscientiously, endeavour to negotiate their parental faith responsibilities with the situation at hand. This can be an emotional and humbling journey. On the one side, many such parents pursue to provide religious justification for continuing end-of-life treatment. Brett and Jersild argue that there are

> four commonly invoked reasons for this: (1) hope for a miracle, (2) refusal to lose hope in God, (3) a conviction that every moment of life is a gift from God and is worth preserving at any cost, and (4) a belief that suffering can have redemptive value.[43]

On the other side, there are undoubtedly equally strong reasons for considering medical treatment to be not as constructive, if

---

[40] Ibidem.

[41] M. al-Nawawī. *Rawḍa al-Tālbīn* [The Garden for Students] vol. 2, p. 98 ('Ālam al-Kutub [The World of Books] Riyadh, Kingdom of Saudi Arabia, 2003).

[42] M. A. Albar, H. Chamsi-Pasha, "Futility of Medical Treatment." F. A. Khan, "The Definition of Death in Islam."

[43] A. S. Brett, P. Jerslid, "'Inappropriate' Treatment Near the End of Life: Conflict Between Religious Convictions and Clinical Judgment," *Archives of Internal Medicine* 163:14 (2003): 1645-9.

administered, as one would desire. In such cases, the family and physicians perhaps aided with the guidance of a religious scholar should objectively consider the particulars of the case in light of religious sensitivities and negotiate a common practical ground based on "the appropriate limits of life-sustaining treatment" rather than a painful and costly process through the legal system where the semantics of the medical discussions may be emotionally distressing to the parents.

Further, the Islamic legal formulations also provide guidance to that effect. The omission of treatment based on the understanding of what is in the "best interests" is neither contra, nor compromises the deeply held Muslim conviction that ultimately all healing emanates from the Divine Being and that no cure is possible except by Divine will and grace.[44] Furthermore, the responsible allocation of finite public resources also needs to be taken into consideration and compared against medical prognosis and potential outcomes of medical treatment. According to Islam, the physician needs to be as certain as possible of the inevitability of the impending death otherwise life should be sustained.[45] Ultimately, the Islamic legal guidance on discharging parental responsibility is that the intervention of medics is viewed as a means to an end rather than the end itself whilst death is inevitable and a passage to eternal life. It should also be noted that whilst absolute cure is in the hands of the Creator and supplication may well be made for cure, this does not necessarily oblige parents to "demand treatment" once contra professional opinion has been received.[46]

IV. Conclusion

This chapter has focused on the emergent debate between clinical judgment aided by modern day technology and religious conviction of terminally ill patients or parents of terminally ill patients. It offers a conceptual framework for discussion between clinicians and relatives to achieve a common understanding on clinical treatment. This dialogue is all the more important when the patient is a minor with a bleak prognosis of survival. To facilitate this dialogue, parental role and responsibilities are discussed and how this role and these

[44] K. A. Choong, M. Chandia, "Technology at the End of Life."
[45] M. A. Albar, H. Chamsi-Pasha, "Futility of Medical Treatment."
[46] Ibidem.

responsibilities can be applied when clinical judgment is against the wishes of parents. This chapter has focused on assessing the common ground on the above from an Islamic perspective. It has drawn on key concepts about life and death and legal maxims from an Islamic viewpoint and has considered the medical decision-making process for a patient in a terminally ill condition. The findings are that a common third ground can be negotiated albeit there may always be parental desire to sustain life irrespective of clinical practice or costs. As technology advances and artificial intelligence is able to provide an earlier and accurate prognosis, negotiating a third space between such ethical dilemmas (clinical judgments against parental convictions), as part of the decision-making-process, may well increase. Only time will tell.

Chapter 9

Do Parents Have a Right to Determine Where a Child Patient Dies?

Lisa Cherkassky
University of Derby, UK

I. Introduction

In a most unusual chapter, it will now be discussed whether parents have a legal right to take their gravely ill children home to die. After four unsuccessful public appeals, there were two additional private hearings for baby Charlie Gard to focus solely on the time and place of his death.[1] Chris Gard and Connie Yates, his parents, tried to control his ending to spend "precious time" with their son but the High Court, after hearing arguments from Great Ormond Street Hospital (GOSH), refused to grant their wishes. On what legal ground might the parents have argued a case for taking their son home to die?[2]

Four potential legal avenues will be explored: (i) the current law on parental autonomy will be studied in the narrow context of death to reveal if there is a right to take a child home to die; (ii) the common law of best interests will also be studied to reveal if a home death could be in the "best interests" of a child; (iii) a new avenue will be explored in "care" law under the Children Act 1989 (if ventilation is removed in a hospital setting first); and (iv) it will be debated whether attempting to

---

[1] On Tuesday, 25th July 2017, and Thursday, 27th July 2017, in the High Court (unreported).

[2] The main legal issue for Alfie Evans was international travel to the Vatican after his brain, due to an unknown progressive disorder, had all but wasted away: *Evans & James v Alder Hey Children's NHS Foundation Trust & Evans* [2018] EWCA 984 (Civ), but for Charlie Gard, the legal fight went one step further to control the time and place of his death. This chapter will therefore focus on Charlie Gard.

control the death of a child is simply a "preference" or "desire" that, owing to its impractical nature, could not be enforced in court.

It will be concluded that notwithstanding a peaceful environment, parents do not have an absolute legal right to control the death of their child, particularly when the plan involves futile treatment at home to "buy more time." However, if palliative care is viewed as "care" rather than "treatment" it may be administered at home under the Children Act 1989 as long as it does not cause significant harm to the child.

II. "A Mother's Promise"

We start from the final public appeal hearing dated Monday 24th July 2017 – *Great Ormond Street Hospital v Yates, Gard & Gard*[3] - whereby Charlie's parents accepted after fresh scans that their son was beyond medical help and discussions turned to taking him home to die.[4] GOSH attended the first of two private hearings the next day to draw up an end-of-life plan for Charlie, and outlined in a position statement the wishes of the parents in clear terms: "Charlie's parents want him to be with them and ventilated at home for several days before receiving palliative care."[5] Their barrister Grant Armstrong described it as "their final wish" to experience "tranquillity outside the hospital" with their son, to have a choice in "the circumstances in which Charlie's passing will be conducted" before he died, and to spend the "maximum

---

[3] [2017] EWHC 1909 (Fam).

[4] For a discussion on the issues raised by the Gard appeals, see: D. Wilkinson and J. Savulescu, "Hard Lessons: Learning From the Charlie Gard Case," *Journal of Medical Ethics,* published online: 2 August 2017, doi: 10.1136/medethics-2017-104492; J. Bridgeman, "Gard and Yates v GOSH, the Guardian and the United Kingdom: Reflections on the Legal Process and the Legal Principles," *Medical Law International* 17:4 (2017): 285-302; E. Cave, E. Nottingham, "Who Knows Best (Interests)? The Case of Charlie Gard," *Medical Law Review* 26:3 (2018): 500; J. Lombard. "Navigating the Decision-Making Framework for Patients in a Minimally Conscious State," *Medico-Legal Journal of Ireland* 22:2, (2016): 78-87; and A. A. Sheikh, "Medico-Legal Issues at The End of Life: Recent Highlights," *Medico-Legal Journal of Ireland* 23:1 (2017): 2-7.

[5] Great Ormond Street Hospital, statement dated 25th July 2017, at paragraph 3. https://www.serjeantsinn.com/news/charlie-gard-position-statements/ (accessed November 3, 2018).

amount of time they have left with Charlie."[6] The High Court was called upon to approve a plan because the hospital would not let Chris and Connie take Charlie home for a further week of futile intensive care, causing talks with palliative care consultants at the hospital to break down.[7] GOSH felt that a dignified death was at risk:

> Charlie is a child who requires highly specialised treatment. His care cannot be simplified. It must be provided in a specialist setting by specialists. It is in Charlie's best interests that the risk of a precipitate, distressing or disordered death is removed so that he may be assured of a peaceful and dignified passing.[8]

GOSH put a suggestion forward - a family-friendly hospice - but time would be limited because it was not licensed or insured to deliver intensive care overnight.[9] Mr Justice Francis gave the parents two days to find an intensive care team for their flat or the hospice who were willing to provide futile intensive care, but hinted that a hospice, with removal of ventilation shortly afterwards, was the realistic option.[10] Chris and Connie were unable to do so. Mr Justice Francis therefore ruled in the final private hearing on Thursday, 27th July 2017, that Charlie was to be moved to a hospice the next day and extubated within a few hours of arrival. Connie Yates expressed her sadness when leaving the High Court:

> [GOSH has] denied us our final wish. We just want some peace with our son – no hospital, no courts, no media – just quality time with Charlie away from everything to say goodbye to him in the most loving way. We've had no control over our son's life and no control over our son's

---

[6] R. Mendick, D. Boyle, "Charlie Gard's Parents Accuse Hospital of Blocking 'Final Wish' for Baby to be Allowed Home to Die," *The Telegraph*, 25th July 2017, https://www.telegraph.co.uk.

[7] See GOSH statement dated 25th July 2017, at paragraph 2, https://www.serjeantsinn.com/news/charlie-gard-position-statements/ (accessed November 3, 2018).

[8] Ibidem, paragraphs 1-5.

[9] Ibidem, paragraphs 8-9.

[10] M. Holden, "Parents, UK Hospital Clash Over Taking Baby Charlie Gard Home to Die," *Daily Mail,* 25th July 2017, https://www.dailymail.co.uk.

> death. I'm shocked that after all we've been through, they won't allow us extra time…we promised Charlie every day we would take him home. It seems really upsetting, after everything we've been through, to deny us this.[11]

GOSH responded in a statement that it had

> tried absolutely everything to accommodate their final wishes…including exploring the unprecedented step of delivering intensive life support away from a hospital intensive care unit [but] there is simply no way that Charlie, a patient with such severe and complex needs, can spend any significant time outside of an intensive care environment safely. The risk of an unplanned and chaotic end to Charlie's life is an unthinkable outcome for all concerned and would rob his parents of precious moments with him.[12]

On the morning of Friday, 28th July 2017, hospice staff transferred Charlie and his parents to a hospice 45 minutes away and they had a few hours to say goodbye. They took him into the garden, made casts of his hands and feet, and took plenty of pictures with him. The ventilator was disconnected at 15:12 PM and he died twelve minutes later. Charlie was given a temperature-assisted "Cuddle Cot" so his parents could take him home (the equivalent of lying in state) for a few days.

On the face of it, the final wish of the parents does seem understandable in the circumstances. So, were they right? Did they have a legal right to take Charlie home to die? The answer is, it depends on

---

[11] Press Association, "Charlie Gard's Parents Prepare for Final Farewell After Hospice Move," *Daily Mail*, 28th July 2017. https://www.dailymail.co.uk; "Charlie Gard's Parents Denied Final Wish After Judge Approves Hospice Plan," *Daily Mail*, 10th August 2017, https://www.dailymail.co.uk; and M. Robinson. "Charlie Gard's Anguished Parents Concede He Must End His Days in Hospice – Yet They Still Cannot Agree with Doctors How He Should be Cared for in his Last Hours," *Daily Mail*, 28th July 2017, https://www.dailymail.co.uk.

[12] Great Ormond Street Hospital, 27th July 2017, https://www.gosh.nhs.uk/news/latest-press-releases/latest-statement-gosh-patient-charlie-gard (accessed November 3, 2018).

whether medical treatment is involved in the plan as to how much control is enjoyed. Chris and Connie did not only wish to take Charlie home to die, but they wished to administer futile intensive care for an additional week in their ground floor flat, without a specialised team, to say goodbye to their son in their own time and on their own terms. This plan caused legal difficulties because of its futility. The four potential legal avenues for parents in this unusual situation are now explored below.

A. Avenue 1: does the existing law on parental autonomy support control over death?

The definition of "parental responsibility" under section 3(1) of the Children Act 1989 is very wide, including: "the rights, duties, powers, responsibilities and authority which by law a parent of a child has in relation to the child and his property."[13] It is also a criminal offence to fail to provide "medical aid" to a child under section 1(2)(a) of the Children and Young Persons Act 1933. The common law regarding parental authority, however, is also clear that even though parents must make decisions about medical treatment, parental authority is subservient to the inherent jurisdiction of the High Court.[14] So, can parental authority be extended to cover the death of a child? There are flickers of support for this idea in the case law. Lord Bingham MR in *Re Z (Identification: Restrictions on Publication)*[15] noted that: "I would accept without reservation that the decision of a devoted and responsible parent should be treated with respect. It should certainly not be disregarded or lightly set aside."[16] A unique scenario occurred in *Re T (A Minor) (Wardship: Medical Treatment)*[17] whereby the parents of a young

---

[13] Parents have a "large measure of autonomy in the way in which they discharge their parental responsibilities" according to *R v Secretary of State for Education and Employment and Others* [2005] UKHL 15, per Baroness Hale of Richmond at para. 72.

[14] *Re A (Children) (Conjoined Twins: Surgical Separation)* [2000] Fam 147 at 178-179; and *Re W (A Minor) (Medical Treatment: Court's Jurisdiction)* [1993] Fam 64 at 78.

[15] [1997] Fam 1.

[16] Ibidem, 32-33.

[17] [1997] 1 WLR 242.

boy refused a liver transplant for their son and the court supported this decision:

> The welfare of this child depends on his mother. She will have to comply with the court order, return to this country and present the child to one of the hospitals. She will have to arrange to remain in this country for the foreseeable future. Will the father stay in country AB and work or come with her to England, giving up his job and having to seek another job? If he does not come she will have to manage unaided. How will the mother cope? I believe that the best interests of this child require that his future treatment should be left in the hands of his devoted parents.[18]

The decision in *Re T* allowed the parents to reject a life-saving liver transplant for their son due to lifestyle, stress, travel, and employment factors. It was also confirmed in *Glass v the United Kingdom*[19] that to treat a child without the consent of the parent interferes with the human rights of the patient (Article 8) and that in the event of a disagreement, court approval is required.[20] *Glass* concerned treatment rather than withdrawal, but it places parents in a position of authority whereby their consent (or lack thereof) is the gateway to medical treatment, rendering doctors powerless until the court intervenes.[21] Parental autonomy is also supported by section 31(2) of the Children Act 1989 whereby a court's jurisdiction is limited to making care or supervision orders if a child is likely to suffer "significant harm" through unreasonable parenting. On the face of it, the ambit of parental responsibility appears to be wide enough to include death.

---

[18] Ibidem, pp. 250-255. For a detailed critical analysis of this case, see: M. Fox, J. McHale, "In Whose Best Interests?" *Modern Law Review* 60 (1997): 700-709.

[19] [2004] 39 EHRR 15.

[20] Ibidem, paragraphs 70 and 75. It was also confirmed in *Lambert v France* [2016] 62 EHRR 2 that court is the best option when doubts arise as to the best interests of the patient (at para. 143).

[21] This was supported in *Re B (A Minor) (Wardship: Sterilisation)* [1988] AC 199; *Re W (A Minor)* [1993] Fam 64; *Re J (Specific Orders: Child's Religious Upbringing and Circumcision)* [2000] 1 FLR 571; and *B (Child)* [2003] EWCA Civ 1148.

However, the question of whether parents have authority over the death of their child ultimately rests on whether medical treatment is part of the plan and the doctors agree with it. Lord Oliver stated in *KD (A Minor) (Ward: Termination of Access)*[22] that: "the natural bond and relationship between parent and child gives rise to universally recognised norms which ought not to be gratuitously interfered with and which, if interfered with at all, ought to be so only if the welfare of the child dictates it."[23] In *Gard*, the court ruled that it did not support the welfare of Charlie to continue his ventilation,[24] meaning that when Chris and Connie proposed to continue the futile intensive care at home, the court was entitled to intervene. The hospital was also under no obligation to support the ventilation once it was rendered futile.[25] It could also be said that the decision in *Re T* was wrongly decided because it placed a greater weight on the lifestyle of the mother than the grave predicament of her son. It has not been followed by any other case since.

The parental autonomy avenue is therefore unavailable to parents wishing to take their child home to die if their wish includes futile treatment: the court can simply override their wish to ensure that the child is spared treatment that is not in his best interests.

B. Avenue 2: can it be in a child's "best interests" to die at home?

A second possibility is whether the best interests test can be interpreted to include a home death. This may offset the futility of the treatment, for example, if there is some form of benefit to be gleaned.[26] This route may be phrased as follows: "is the continuation of treatment at home

---

[22] [1988] AC 806.

[23] Ibidem, pp. 824-825.

[24] *Great Ormond Street Hospital v Yates, Gard & Gard* [2017] EWHC 972 (Fam).

[25] Indeed, the case of *Portsmouth NHS Trust v W* [2005] EWHC 2293 confirms that a doctor need not act if it is against his "professional conscience, intuition or hunch" per Hedley J at para. 56, following on from Lord Donaldson MR in two cases: *Re J (A Minor) (Wardship: Medical Treatment)* [1991] Fam 33 at 41; and *Re J (A Minor) (Child in Care: Medical Treatment)* [1992] 3 WLR 507 at 516.

[26] The best interests test is only applicable if medical treatment is involved in the home death, so if the parents agree to withdrawal in a hospital and *then* take their child home to die, there may be no need for a best interests test. This option will be discussed below as "avenue 3."

with the ultimate aim of ending life surrounded by family in the best interests of the child?" The case law on best interests supports a comprehensive test which could include a home death. For example, the case of *Re A (Male Sterilisation)*[27] interpreted "best interests" to include "medical, emotional, and all other welfare issues."[28] *An NHS Trust v MB*[29] also interpreted best interests "in the widest sense" to form a balance sheet of benefits and burdens, including "medical, emotional, sensory and instinctive considerations."[30] *Aintree University Hospital NHS Trust v James*[31] added that: "decision makers must look at welfare in the widest sense, not just medical but social and psychological."[32] These judgments support the idea that a gravely ill child could die at home surrounded by family if this provides him with an emotional benefit.

However, should that child be unconscious and any emotional or physical benefit be removed, the question then boils down to simply: "is the continuation of futile treatment in the best interests of the child?" The cases cited above are not so supportive when read in the context of futility. *Re A*, for example, stated that "speculative benefits" are not part of the best interests test,[33] and in *Gard* where the court confirmed that there was "nothing to be put on the other side of the balance in favour of preserving life,"[34] it is highly unlikely that a child would glean any (even a speculative) benefit from being taken home for futile treatment so as to fulfil the desires of his parents. The court in *MB* also rejected the "wholly irrelevant" wishes of the parents, meaning that the home death would have to confer an actual benefit to the child to be considered.[35] The court in *Portsmouth NHS Trust v Wyatt*[36] raised

[27] [2000] 1 FLR 549.

[28] Ibidem, per Dame Elizabeth Butler-Sloss P at p. 555.

[29] [2006] EWHC 507.

[30] Ibidem, per Holman J at para. 16. See also: *An NHS Trust v A* [2007] EWHC 1696, per Holman J at para. 40.

[31] [2013] UKSC 67.

[32] Ibidem, per Lady Hale at para. 39.

[33] *Re A (Male Sterilisation)* (n 27) per Dame Elizabeth Butler-Sloss P at p. 555.

[34] *Yates & Gard v Great Ormond Street Hospital* [2017] EWCA Civ 410, 23rd May 2017, per Lord Justice McFarlane at para. 48.

[35] *NHS Trust v MB* [2006] EWHC 507, per Holman J. at para. 16.

[36] [2004] EWHC 2247 (Fam).

concerns that the parents were "projecting [their] intuitive feelings" onto their gravely ill daughter and found that prolonging her life to allow her parents to control her death was not in her best interests.[37] *Re L (A Child) (Medical Treatment: Benefit)*[38] – with remarkably similar facts to *Gard* – also determined that even though the term "medical interests" included the emotional need of the baby to bond with his mother, the medical treatment was futile, rendering the balance sheet in favour of withdrawal.[39] Therefore, the best interests test when applied to a home death leads to the same negative result as seen in *Gard* if medical treatment is involved: the continuance of futile treatment will not be in the best interests of the child. The only benefit is to the parents. It is a well-established principle going back to *Re J (A Minor) (Wardship: Medical Treatment)*[40] that futile treatment must be withdrawn, as per Lord Donaldson MR: "in the end there will be cases in which the answer must be that it is not in the interests of the child to subject it to treatment which will cause increased suffering and produce no commensurate benefit."[41]

It is worth noting that substituted judgment is also not an option. Desperate parents may use phrases such as: "he would have wanted it" or "this would have been his wish if he were awake" to substitute the consent of the child, but the idea of a substituted judgment was removed from minors in *Airedale NHS Trust v Bland*[42] and removed from incompetent adults in *Re F (Mental Patient: Sterilisation)*.[43] A best interests test is now deemed more appropriate.[44] There is a glimpse of substitut-

[37] Per Hedley J at para. 34. For further discussion, see: M. Brazier, "An Intractable Dispute: When Parents and Professionals Disagree," *Medical Law Review* 13:3 (2005): 412.

[38] [2004] EWHC 2713.

[39] Ibidem, per Dame Elizabeth Butler-Sloss P at para. 26.

[40] [1991] Fam. 33.

[41] Ibidem, p. 47.

[42] [1993] AC 789, per Lord Goff and Lord Mustill at pages 871-872 and 894-895 respectively.

[43] [1990] 2 AC 1, per Lord Brandon at 55.

[44] For further discussion on the inappropriateness of substituted judgment on minors, see: L. Cherkassky, "Children and the Doctrine of Substituted Judgment," *Medical Law International* 1 (2015): 1-23; D. Tomkin, P. Hanafin, "Medical Treatment at Life's End: The Need for Legislation," *Medico-Legal Journal of Ireland* 1:1 (1995): 3; and C. Somers, "Deciding Obliquely and by a Side Wind:

ed judgment in *Aintree* whereby Lady Hale said: "[the doctors] must try and put themselves in the place of the individual patient and ask what his attitude to the treatment is or would be likely to be,"[45] but this was an incompetent adult case whereas the statutory best interests test under section 4(6) of the Mental Capacity Act 2005 requires the medical team to ascertain the wishes and feelings of the patient during his competent years. It does not substitute his judgment - it tries to ascertain his actual judgment - and so it would not be applicable to children who had not yet lived a full life. The best interests' avenue is therefore closed to parents if futile treatment is involved in the home death.

C. Avenue 3: a new approach under Care Law

Let us now remove futile medical treatment from the equation and simply pose the legal question as follows: "do parents have a right to take their gravely ill child home to die?" The answer appears to be "yes."

Parents who agree to withdraw ventilation at the hospital before taking their child home remove any medical element from the equation, leaving the Children Act 1989 to govern their parental responsibility under "care" law. This requires a welfare test under section 1(1) which stipulates: "when a court determines any question with respect to the upbringing of a child, the child's welfare shall be the court's paramount consideration," but a court can only intervene and apply this test if there is a risk of "significant harm" to the child under section 31(2), which states:

> A court may only make a care order or supervision order if it is satisfied that the child concerned is suffering, or is likely to suffer, significant harm and that the harm, or likelihood of harm, is attributable to the care given to the child, or likely to be given to him if the order were not

Substituted Judgment and End-of-life Decisions for Minors," *Medico-Legal Journal of Ireland* 19:1 (2013): 11.

[45] *Aintree University Hospital NHS Trust* (n 31), para. 39.

> made, not being what it would be reasonable to expect a parent to give to him.[46]

A home death has not yet been debated within the context of care law, nor has a court yet ruled on it, but perhaps it may offer a legal avenue for future cases like Charlie Gard?[47]

Chris Gard and Connie Yates did try to exert their parental authority using section 31(2) to argue that taking Charlie abroad for innovative treatment did not breach the threshold of significant harm and therefore the courts had no place to intervene. This was successful in the earlier case of *Re King*[48] whereby Baker J said:

> it is a fundamental principle of family law in this jurisdiction that responsibility for making decisions about a child rest with his parents. In most cases, the parents are the best people to make decisions about a child and the State – whether it be the court, or any other public authority – has no business interfering with the exercise of parental responsibility unless the child is suffering or is likely to suffer significant harm as a result of the care given to the child not being what it would be reasonable to expect a parent to give.[49]

Nevertheless, Lord Justice McFarlane in the Court of Appeal in *Gard*[50] quickly rejected the idea that care law could govern medical treatment and restated the importance of the common law of best interests:

> Best interests is the established yardstick which applies to all cases...Mr Justice Baker's words provide no basis for saying that he was holding that any test based on significant

---

46 For further discussion, see: C. Auckland and I. Goold, "Defining the Limits of Parental Autonomy: Charlie Gard, Best Interests and Risk of Significant Harm Threshold," *Law Quarterly Review* 134 (2018): 37-42.

47 It should be noted, however, that pain relief (palliative care) may be considered medical treatment, so there might still be medical interference with parental autonomy. This is unclear without existing case law.

48 [2014] EWHC 2964.

49 Ibidem, paragraphs 31 and 34.

50 *Yates & Gard v Great Ormond Street Hospital* [2017] EWCA Civ 410.

> harm is to be applied to cases relating to the medical treatment of children.[51]

However, should Chris and Connie have agreed to withdraw ventilation at the hospital and *then* take Charlie home to die, the need for a best interests test would have dissolved, paving the way for the 1989 Act. The legal question would then be rephrased as follows: does taking Charlie home to die carry a risk of significant harm under section 31 of the 1989 Act? One can only imagine that, with adequate pain relief, and notwithstanding his inevitable natural death, the answer would be "no."

There may be a potential barrier to this legal avenue. No common law exists at present, but a local authority may disagree with the parents that taking their child home to die will not cause him significant harm. This impasse would result in the High Court invoking its inherent jurisdiction to rule on the matter under section 100(4) of the 1989 Act, which states that the court must have "reasonable cause to believe" that if its inherent jurisdiction is not exercised, a child will suffer "significant harm." The courts may find it harder to overrule parents in a "death only" scenario if the parents have already agreed to withdraw futile treatment in the hospital.

The recent case of *Re C (Children)*[52] demonstrates how the law works in practice and how the courts might deal with a home death scenario. A mother wished to name her daughter "Cyanide," so the local authority sought the inherent jurisdiction of the High Court under section 100(4) of the 1989 Act. The three legal questions to be answered at appeal were: (i) whether the local authority had the statutory power to prevent the parents from naming their children; (ii) whether the local authority, if they did have this power, needed to ask the court to sanction its authority, and (iii) whether the naming of children was serious enough to come under the inherent jurisdiction of the court. All three questions were answered affirmatively, allowing the court to explain in detail why certain "unusual, bizarre, extreme or foolish" parents required court intervention.[53]

---

[51] Ibidem, paragraphs 74 and 104.

[52] [2016] EWCA Civ 374.

[53] Ibidem, per King LJ, para. 105.

In answer to the first question, the local authority did have the power under section 33(3)(b) of the 1989 Act to determine how parents meet their parental responsibility, which states:

> while a care order is in force with respect to a child, the local authority designated by the order shall: (a) have parental responsibility for the child, and (b) have the power to determine the extent to which [a parent] may meet his parental responsibility for him."

Lady Justice King described this provision as a "trump card" allowing a local authority to intervene with the "rights, duties, powers, responsibilities and authority" of parents under section 3(1).[54] There are limitations to this power listed under sections 33(6) and (7) whereby the local authority cannot act without a court order (such as taking a child abroad) but nothing is listed about forenames, allowing the local authority in *Re C* to: "exercise its parental responsibility under section 33(3) in order to prevent the mother from giving her twins the forenames of her choice."[55] A home death would come under the ambit of section 33(3)(b) because it is not carved out as an exception under sections 33(6) and (7) requiring a court order. The local authority has the statutory authority to intervene with preventative measures.

In answer to the second question, the local authority was right to ask the court to sanction its statutory power owing to the potential breach of the human rights of the parents (per Lady Justice King):

> the mother has a safety net in that she may apply for an injunction under section 8 of the Human Rights Act 1998 … such a decision involves such a serious invasion of the Article 8 rights of the mother that I am satisfied that the court should invoke its inherent jurisdiction in order that it may either sanction the local authority's proposed course of action as in the interests of the child.[56]

---

[54] Ibidem, paragraphs 59 and 64.

[55] Ibidem, per King LJ, paragraphs 61, 65 and 66.

[56] Ibidem, per King LJ, paragraphs 75, 77 and 98. King LJ also confirmed that the court was there to "limit, circumscribe or sanction" the existing power of the local authority under section 33(3)(b), not "confer" the power onto the local authority (at para. 97).

A home death is also a private and family matter, so an action by the local authority to prevent it may constitute an unjustifiable interference of the right to respect for the private and family life of the family (both the parents and the child) under Article 8 of the European Convention of Human Rights.[57] The local authority, therefore, notwithstanding its statutory powers under section 33(3)(b) to prevent the home death to avoid significant harm to the child, may still have to invoke the inherent jurisdiction of the court to sanction its preventative measures. It would then be for the court to decide whether there is an unjustified interference under Article 8 and whether there is a risk of significant harm to the child. There is no indication as of yet as to how the court would decide this matter.

In answer to the third question, and notwithstanding the potential human rights breach, naming a child "Cyanide" was deemed by Lady Justice King to be serious enough to invoke the inherent jurisdiction of the court. The "significant harm" test under section 31(2) was applied at this point to answer the pivotal question posed by section 100(4): "whether there is reasonable cause to believe that if the court's inherent jurisdiction is not exercised with respect to the child [she] is likely to suffer significant harm."[58] Lady Justice King concluded that:

> in my judgment, although it will only rarely be the case, there is every reason to believe that if the court's inherent jurisdiction is not invoked in order to prevent the girl child from being named Cyanide, she is likely to suffer significant harm.[59]

King LJ then carved out a "special category" of cases for future reference that could be applicable to home deaths:

> I have reached the conclusion that there is a small category of cases where, notwithstanding the local authority's powers under section 33(3)(b), the consequences of the exercise of a particular act of parental responsibility are so profound and have such an impact on either the child or

[57] Ibidem, per King LJ, para. 75.
[58] Ibidem, per King LJ, para. 101.
[59] Ibidem, para. 102.

> the Article 8 rights of those other parties, that the matter must come before the court for its determination.[60]

It is highly probable in a home death scenario that taking a child home to die falls within the "special category of cases" that requires court approval. This is because, first, the child may suffer intense pain or distress once his life-sustaining treatment has been removed. Secondly and conversely, the consequences of not letting the parents do this would almost certainly interfere with their rights under Article 8.[61] It is also fair to say that, if the name "Cyanide" is deemed by the courts to cause significant harm, then taking a gravely ill child out of hospital will almost certainly attract the attention of the law.

The first home death case, therefore, may decide the following: (i) the local authority has the statutory power to prevent parents from taking their child home to die under section 33(3)(b) because it is not listed as an exception requiring a court order under sections 33(6) and (7) of the 1989 Act; (ii) the local authority will need to ask the court to sanction its preventative measures because a home death falls into a special category of cases whereby the human rights of numerous parties may be interfered with; and (iii) a home death is serious enough to fall within the inherent jurisdiction of the court because of the risk of significant harm to the child. It is very difficult to predict how the court will answer this final and pivotal question, but the complexity of the grave illness and the level of pain relief will be decisive factors. In favour of the parents, palliative care is easier to administer at home than intensive care, and once the medical treatment of a child ceases, the hospital has no legal grounds upon which to further detain the patient. This would be a deprivation of his liberty under Article 5 (a theory also unexplored by the courts in the context of home death thus far).[62]

---

[60] Ibidem, paragraph 104.

[61] It was confirmed in *Lambert v France* [2016] 62 EHRR 2 that relatives who were also victims of a human rights breach may have scope for an action under human rights law.

[62] It should be noted, however, that this section is underpinned by the assumption that "medical treatment" stops when ventilation is withdrawn. If palliative care (pain relief) is considered to be medical treatment too, we have a completely different scenario: we return back to the "doctors versus parents" scenario under the common law of best interests as per the *Gard* case. There is no

D. Avenue 4: it is not a legal right but a fanciful whim dependent on practicalities

Chris Gard and Connie Yates asked to take Charlie home for an extra week of futile intensive care before withdrawing his ventilation. This idea has now been explored and it is clear that parents do not have a legal right to continue futile medical treatment in order to create an "ideal" death for their child. The cases of *Re King*[63] and *Re C (Children)*[64] presented a more viable option, which is to take a child home to die *after* futile treatment has been withdrawn, but this may only be viable if the doctors are supportive of palliative care (pain relief) in a private dwelling under "care" law (the Children Act 1989). The courts have not ruled on this matter yet, but should a disagreement arise in future and the courts agree with the doctors that palliative care is to be administered in hospital to avoid significant harm to the child, the parents have no legal avenue left except to plead that it is their "wish" to take their child home.

How unreasonable is this wish? Two days after the European Court rejected their appeal, Chris and Connie posted devastated video messages on YouTube describing how they felt unable to control their son's death (on Thursday 29th of June 2017):

> We're not allowed to choose if our son lives and we're not allowed to choose when or where Charlie dies. We know what day our son is going to die but don't get a say in how that will happen.[65]

The public did not feel that Chris and Connie were being unreasonable. On the morning of Mr Justice Francis' decision to send Charlie to

legal authority yet on whether palliative care is regarded as "treatment" or simply "care."

[63] [2014] EWHC 2964.

[64] [2016] EWCA Civ 374.

[65] D. Boyle, "Charlie Gard's Parents Endure 'Worst Day of Our Lives' as Doctors to Switch Off Baby's Life Support Within Hours," *The Telegraph*, 30th June 2017. https://www.telegraph.co.uk; and M. Robinson, "Outpouring of Grief Worldwide for Baby Charlie Gard as his Parents say their Last Goodbyes to their Son as Doctors Switch Off his Life Support after Eight-month Battle," *Daily Mail*, 30th June 2017. https://www.dailymail.co.uk.

a hospice, a family friend of Connie Yates posted on Facebook (on Thursday, 27th July 2017):

> the hospital has set the bar so high that in terms of [a] clinical team for Charlie's end of life nothing seemed good enough for Great Ormond Street … Connie and Chris have conceded a hospice but it was not their first choice. They will be devastated they have not been granted their final wishes as parents.[66]

This message reflected a wider sense of public anger that Chris and Connie's final wishes were not respected. However, the public misunderstood three vital issues. First, it was simply not practical to administer futile intensive care in a ground floor flat. GOSH provided a number of reasons for this, including bulky equipment, no qualified experts, and the risk of an "unplanned and chaotic" death which would have been "unthinkable" for the parents.

Secondly, and most importantly, the courts had already ruled that Charlie's ventilation was futile, so it was not appropriate to continue it any further. It is here that the motives of the parents changed too: when they conceded on Monday, 24th July 2017,[67] that Charlie was beyond medical help, their priorities switched from what was "best for him" to what was "preferable for us," indicating that the best interests of Charlie had expired.

Thirdly, and in a general moral sense, parents cannot control the natural death of their child. They can try to prevent death with proper parenting (e.g. road safety or good diet), and they can try to chase away the threat of death with medical treatment (e.g. chemotherapy), but when these options fail, and natural death is imminent, no parent has the luxury of asking for "one more week." The crux of the two additional private hearings on Tuesday, 25th July 2017, and Thursday, 27th July 2017, was for Chris and Connie to control the time and location of Charlie's death, but parents who tragically lose their children to grave genetic diseases do not have this luxury. Chris and Connie were ulti-

---

[66] Press Association, "Anguish of Charlie Gard's Mother as Judge Sets Timetable for End of his Life," *Daily Mail,* 27th July 2017. https://www.dailymail.co.uk.

[67] In *Great Ormond Street Hospital v Yates, Gard & Gard* [2017] EWHC 1909 (Fam).

mately playing God when a hand had already been dealt. They were attempting to postpone his natural death. The "right" so revered by the public would be better termed as a "preference" or a "desire" that is ultimately relegated by nature.

III. Conclusion

The answer to the opening question: "do parents have a right to determine where a child patient dies?" is that it depends. A request to continue futile intensive care at home (as per Charlie Gard) will not be in the best interests of the child. However, palliative care may be administered at home under the "care" provisions of the Children Act 1989 if it is not considered to be "treatment" (the courts have not yet ruled on this matter but *Re King* and *Re C* provide good guidance as to how the "significant harm" test would apply). Ultimately, however, it is still not an absolute legal right as much as it is a "wish" that is subservient to nature taking its own course, the seriousness of the illness, and the complexity of pain relief required.

There is one final theory to be considered: perhaps Chris and Connie attempted to control the death of their son because their control over his *life* had been removed? Was it a way of "repossessing" their own child? Their frustration was clear after the Court of Appeal ruling (Chris Gard): "When we got the appeal papers they said Connie Yates and Chris Gard against Great Ormond Street Hospital and Charlie Gard and yet he's our son. It broke my heart when I saw that because how can that be right?"[68] Connie published an interview after Charlie had died, and the peaceful way she describes those few days at home suggests that all they really wanted was to be his parents again:

> Once home, it was lovely to sit and watch him, lying there like any other baby, not surrounded by equipment and machinery, without anything obscuring his lovely face. To just see our Charlie, at home, sleeping in his cot where he should be … it felt perfectly natural to leave the hospice with Charlie and take him home with us. We had got our last wish to bring him home, but Charlie was no longer

[68] S. Bell, "Charlie Gard Parents to Keep Fighting," *BBC News Online*, 1st June 2017. https://www.bbc.co.uk/news.

> alive. Although it was upsetting, at least he was home – finally back where he belonged. It felt like he was ours again.[69]

It is submitted that, if the myriad of legal actions and the adversarial legal documents had been replaced by non-technical mediation (e.g. no medical jargon or intimidating specialists) and a stronger emphasis was placed on the option of taking Charlie home to lie in state for a few days, perhaps Chris and Connie would have been more open to the idea of letting him die in a hospital or hospice, and would have had a glimmer of peace to look forward to in a world of chaotic darkness.

[69] A. Smith-Squire, "Our Last Hours With Our Son: Charlie Gard's Parents Emotionally Reveal How They Finally Brought Their Baby Home After he Died in a Hospice and Spent Several Days Saying Their Last Goodbye," *Daily Mail,* 4th August 2017. https://www.dailymail.co.uk.

PART II

International Perspectives

Chapter 10

Serving the Child's "Best Interests" in Australia

Roslyn Jones
Independent Scholar, Australia

I. Introduction

Medical advancements allow the treatment of many conditions that once ended in certain death. We also have the ability to maintain circulation and ventilation even after the heart has stopped beating or the brain has stopped functioning. There has been a consequent shift in how we view the concept of death, and how we view omissions or withdrawals of medical interventions that lead to a hastening of death or an allowance of nature to take its course.[1]

This shift is often most poignant in the plight of the newborn and young. There can be few clinical decisions that are more difficult than those that involve determining whether a medical treatment is futile with the consequent decision to withhold treatment when it involves the life of a baby or a young child. Conversely, it is distressing for clinicians to be involved in situations when parents oppose their advice for potentially life-saving treatment for a child. These are complex situations, the two cases experienced in the UK of Charlie Gard and Alfie Evans (henceforth referred to as Charlie and Alfie) being recent high profile examples that illustrate this clearly.[2]

In Australia, consensus with parents on treatment for babies and children is generally reached by establishing strong communication channels between the parents and the treating team. Decisions are ide-

---

[1] C. Stewart, "Death Management in the Case of Baby D," *ABC Religion and Ethics* (published on May 3, 2011), https://www.abc.net.au/religion/death-management-and-the-case-of-baby-d/10101488 (accessed July 23, 2018).

[2] See the Introductory chapter of this book.

ally reached in a situation of support and trust, on the basis of information on prognosis and treatment options.

However, this approach is not always successful and in times of disagreement, other options are explored in an attempt to reach a resolution. These include: obtaining a second opinion; seeking third party mediation by involving ethics committees; and by providing counselling or pastoral care. When these efforts are not successful, other means include the arrangement of the transfer of care; treating according to the parents' wishes, at least for a period of time; treating according to the parents' wishes while legal proceedings are commenced; or treating according to the clinician's advice and leaving the responsibility of arranging legal proceedings to the parents.[3]

So, withholding treatment cases do present before the court on occasion. The legal background on which these cases are heard will be explained, and four cases will be initially outlined and then analysed with respect to anticipating Australia's response to the controversial UK cases of Charlie and Alfie.

II. Legal foundations in Australia

The foundations upon which legal determinations are based in these cases are those of the "inherited" principles of law from England and Wales. Common law migrated with the convicts on settlement in 1788. With the Australia Courts Act 1828, the *parens patriae* jurisdiction of the English Court of Chancery became incorporated into Australian law.[4] This jurisdiction of the superior courts is Latin for "parent of the nation" and derives from the royal prerogative to protect the person and property of those subjects not legally competent to attend to their own

[3] R. Drake et al, "Decision-Making at the End of Life in Infants, Children and Adolescents," *The Royal Australasian College of Physicians, Paediatrics & Child Health Division* 16 (2018).
https://www.racp.edu.au/docs/default-source/advocacy-library/decision-making-at-the-end-of-life-in-infants-children-and-adolescents.pdf (accessed November 17, 2018).

[4] P. L. G. Brereton, "The Origins and Evolution of The Parens Patriae Jurisdiction," The Honourable Justice Paul L. G. Brereton AM RFD Lecture on Legal History Sydney Law School, on Friday, May 5, 2017.
http://www.supremecourt.justice.nsw.gov.au/Documents/Publications/Speeches/2017%20Speeches/Brereton_050517.pdf (accessed November 14, 2018).

affairs.[5] The Supreme Court of each state and territory in Australia has the power of *parens patriae*. The overriding ethos of such legislation is to secure the welfare of children and to define the circumstances by which they are at risk of neglect or abuse.[6]

In the mid-1880s, there was also a growing attention to child protection within society, which led to the introduction of children's courts in the early 1900s, the intention being to advocate on the behalf of children and to ward against cruelty and child offence. By the 1970s, there was federal in addition to state and territory legislation, in which the reasoning of "the best interests of the child" was adopted.[7] The state, as *parens patriae*, and the family, intersect under this legislation in this arena of child protection. As will be explained, this legislation extends not only to situations of abuse or neglect, but also to the protection of the "person" in general terms, which naturally includes medical treatment.

In Australia, parents generally have the lawful authority to consent to medical treatment for infants and young persons.[8] This right is based on the parental role as guardian of the child until the child reaches the age of 18, at which time he or she can consent as an adult. However, this authority is only unfettered if the parents are acting in the "best interests" of the child, and any person who is concerned about the

---

[5] C. Smith, "Parens Patriae Jurisdiction," presentation to Legal Aid Queensland on July 24, 2016, https://elo.legalaidqld.gov.au/webdocs/internal/irregseries/cle/2018/parens.pdf (accessed October 26, 2018).

[6] Australian Law Reform Commission, "Purposes of Laws Relevant to Family Violence, Child Protection Law," https://www.alrc.gov.au/publications/4.PurposesofLawsRelevanttoFamilyViolence/child-protection-law (accessed November 14, 2018).

[7] Ibidem.

[8] Under the Family Law Act, Section 63E "(1) A person who is the guardian of a child under this Act has responsibility for the long-term welfare of the child and has, in relation to the child, all the powers, rights and duties that are, apart from this Act, vested by law or custom in the guardian of a child, other than: (a) the right to have the daily care and control of the child; and (b) the right and responsibility to make decisions concerning the daily care and control of the child. (2) A person who has or is granted custody of a child under this Act has: (a) the right to have the daily care and control of the child; and (b) the right and responsibility to make decisions concerning the daily care and control of the child."

child's welfare or treatment can challenge the decisions of the parents by applying to the court on the grounds that a parental decision is not in the child's best interests. This is on the basis of the aforementioned power of the court in being able to supervise parents and other guardians[9] in the court's role as the protector of the welfare of children in the interests of society.

The jurisdiction of *parens patriae* can be far reaching in the application of this purpose. Its power is more extensive than that of parents in that "it can be invoked in such matters as custody, protection of property, health problems, religious upbringing and protection against harmful associations."[10] Parents cannot consent to certain interventions for their children such as sterilisation, involuntary admission (for instance for conditions such as anorexia nervosa), the donation of a child's bone marrow, treatment for intersex, treatments for gender identity disorder, or indefinite protective detention.[11]

But there is also restraint in the application of *parens patriae.* This power is to be exercised only in exceptional cases and used with particular caution in those matters where there is "some clear justification for a court's intervention to set aside the primary parental responsibility for attending to the welfare of the child."

Exceptional cases can and do arise under the jurisdiction of *parens patriae* with respect to medical interventions for a child as a result of applications to the court from either the parents or from treating practitioners, regarding treatment that is to be withheld or withdrawn or to be administered against the wishes of the parents. The legal test for assessment of a child's "best interests" in these cases includes a consideration of the physical effects of the proposed treatment together with any psychological and social implications of this intervention.[12]

It is important to understand that these cases are considered in the broader contexts of both the Australian public health system and the common law legal framework. Australia has a federally funded universal health coverage system (Medicare) and maintains a strong public hospi-

[9] P. L. G. Brereton, "The Origins," 11.

[10] Ibidem, 8.

[11] C. Stewart, "Death Management."

[12] S. Bird, "Mohammed's Case," *Australian Family Physician* 43:5 (2014): 332. https://www.racgp.org.au/afp/2014/may/mohammeds-case/ (accessed November 14, 2018).

tal system that is state managed. Medical practitioners are registered under a national system that regulates qualifications and practice, and hospitals comply with state health legislation as well as ministerial health policies.

In addition to these levels of control, there are extensive common law provisions that are relevant to medical practitioners. Significantly, under these provisions, medical practitioners are not obliged to provide medical treatment that is deemed to be "futile." The basis for the determination of futility is that the treatment is of no benefit to the patient; or that the burdens of treatment are out of all proportion to any potential benefit of that treatment for the patient. Such a medical assessment is, by preference, made in accordance with clinical guidelines and/or hospital policies.[13]

III. Case law

There is a range of Australian case law that can be cited to illustrate these principles in action, both with respect to enforcing and withholding treatment in minors. These former cases often involve the administration of blood transfusions when there is a religious objection, most notably from Jehovah's Witnesses.

However, for these purposes, four cases will be outlined in detail:

- *Secretary, Department of Health and Community Services v JWB & SMB (Marion's case)*[14]
- *Baby D (No. 2)*[15]
- *TS & DS v Sydney Children's Hospital Network (Mohammed's case)*[16]
- *Director of Clinical Services, Child & Adolescent Services v Kizsko* [17]

These cases have been selected in order firstly to provide the foundation of these principles (especially "Marion's case") and secondly to serve as a basis for comparison with the matters of Charlie Gard and Alfie Evans. What follows is a brief summary of the cases.

---

[13] Ibidem.

[14] [1992] HCA 15; (1992) 175 CLR 218. *Re Marion (No 2)* [1992]17 Fam CA 87 (1 May 1992).

[15] [2011] FamCA 176 (16 March 2011).

[16] [2012] NSWSC 1609.

[17] [2016] FCWA 19, FCWA 34, FCWA 75.

A. Secretary, Department of Health and Community Services v JWB & SMB (Marion's case) and Re Marion (No. 2)

One often-quoted case is that of the so-called "Marion's case" which commenced in 1990 before the Family Court of Northern Territory but continued on appeal before the High Court of Australia in 1991. This case dealt with issues involving the sterilisation of an intellectually disabled girl. Marion's parents had asked the Court to either grant permission for the sterilisation or to deem it lawful for parents to consent to the procedure. The majority of the High Court found that common law does not extend rights for parents to consent for non-therapeutic medical procedures such as sterilisation which was, in this instance, indicated on social rather than medical grounds. Instead, the High Court determined that the Family Court of Australia had jurisdiction to authorise the procedure, which it subsequently did at the 1991 hearing.

This case was significant in that it introduced the need for court authorisation for certain procedures or interventions for children. Further, Nicholson CJ at the subsequent Family Court hearing catalogued a number of factors that were important in any decisions concerning orders in relation to medical procedures. He listed these as follows, which is worth quoting in full in order to convey the relevant issues:

> (i) the particular condition of the child which requires the procedure or treatment;
> (ii) the nature of the procedure or treatment proposed;
> (iii) the reasons for which it is proposed that the procedure or treatment is to be carried out;
> (iv) the alternative courses of treatment that are available in relation to that condition;
> (v) the desirability and effect of authorising the procedure or treatment proposed rather than the available alternatives;
> (vi) the physical effects on the child and the psychological and social implications for the child of:
>
> (a) authorising the proposed procedure or treatment
> (b) not authorising the proposed procedure or treatment
>
> (vii) the nature and degree of any risk to the child of:
>
> (a) authorising the proposed procedure or treatment

(b) not authorising the proposed procedure or treatment

(viii) the views (if any) expressed by:

(a) the guardian(s) of the child

(b) a person who is entitled to the custody of the child

(c) a person who is responsible for the daily care and control of the child

(d) the child to the proposed procedure or treatment and to any alternative procedure or treatment.[18]

B. Baby D (No 2)

This catalogue was relevant to a case heard in Victoria in 2011 concerning "Baby D" who was born at 27 weeks gestation. This was a twin pregnancy and both neonates required resuscitation post-delivery. The twin of Baby D progressed well, but the course of Baby D was more complicated, with brain damage subsequent to extubation attempts. She was reintubated and fed via a nasogastric tube. The baby responded to touch, could feel pain and was often distressed. She still had brainstem function. At five months, it was considered that the ventilation tube should be removed but there was uncertainty about whether this should be replaced if there was deterioration in the child as a result.

The matter went before the hospital's ethics committee. The consensus of the committee was to remove the tube, and should there be respiratory distress as a result, it would be appropriate for the tube not to be replaced but the baby to be given palliation to reduce pain and suffering. However, the committee was hesitant to recommend this course of action and advised that the advice of the Family Court be obtained.

The issues before the court were then:

- firstly, whether the parents would be able to authorise and consent to the extubation of the baby;
- secondly, whether they could authorise the palliation of the baby should there be consequent pain and/or respiratory distress following extubation. Such palliation would include the

[18] *Department of Health & Community Services* (n 14), para. 5.

administration of sedation or other medication as deemed necessary by medical staff; and

- thirdly, should such consent be given, whether medical staff could withhold treatment that would artificially prolong the life of Baby D.

The judge hearing the case concluded that the parents were authorised to give consent to extubation and palliation and that the medical staff was permitted to withhold treatment. He found that the treatment decision reached by the parents in consultation with the doctors was appropriate and that no criminal sanctions would prevail should death result. The judge summarised the evidence as follows:

> That eventual acceptance of her death is no doubt both difficult and distressing to parents and all professionals but at that stage the very best of effort and medical knowledge has been available to Baby D and there is a point in time where her best interests do require efforts to ease her distress with the knowledge and acceptance of her death. [19]
>
> The consensus amongst all of the very experienced and qualified medical practitioners … is that any future for Baby D must, with certainty, be seen to be one that is, at least, very burdensome and futile with no expectation of any enjoyment of life and without sight and any meaningful brain capacity.[20]

The judge also found that the doctors caring for Baby D may have been unable to relieve her pain and distress without the use of palliation, despite the risk of such medication further depressing the baby's breathing once the tube was removed. He concluded that the administration of palliative drugs was humane.

Significantly, the judgment was that it was in the baby's best interests to administer the medication even if it shortened her life, a reference to the ethical principle of the Catholic church of "the doctrine of double effect." This principle allows medications that are used primarily to relieve pain in a terminal situation, despite the unintended effect of the medications shortening life.

[19] *Baby* D (No. 2) (n 15), para. 148.

[20] Ibidem, para. 149.

So this case illustrates the ideal of parents and doctors consulting in what is in the child's best interest. But it also demonstrates that even with parental, medical and ethics committee accord, there was still some uncertainty about the "lawfulness" of suppressing a drive to breathe should the baby develop distress from the proposed extubation.[21] The judge's determination allowed the parents to give consent to proceeding with a medical intervention that would likely lead to end-of-life care.

C. TS & DS v Sydney Children's Hospital Network (Mohammed's case)

A further case in 2012 was heard by the Supreme Court of New South Wales when the parents of a nine month old infant, who suffered from a mitochondrial disorder, made an application to the Supreme Court seeking an order to ensure that their child be kept alive by mechanical ventilation. Mohammed was deaf, blind, and unable to move. He suffered from seizures, needed to be artificially fed, and only responded to painful stimuli.

In response to the parents' application, an urgent sitting was convened at the hospital and evidence was heard from the parents, a treating paediatrician and two paediatric intensivists. The questions as defined by the judge were: was it in Mohammed's best interests to be mechanically ventilated, and if it was in Mohammed's best interests, should the Court order mechanical ventilation contrary to the treating doctors' recommendations?[22]

The judge in this matter agreed with the clinicians that it was not in Mohammed's best interests to sustain mechanical ventilation when his condition was incurable and when there was considerable pain and distress for the child associated with the proposed ventilation. He concluded:

> Here I am well satisfied that the doctors' opinions as to Mohammed's best interests have been reached conscien-

---

[21] Ibidem, para. 81.

[22] *TS & DS* (n 16), para. 88.

tiously and in the proper discharge of their professional obligations.[23]

D. Director of Clinical Services, Child & Adolescent Services v Kizsko

A more recent case, about which there were three hearings heard in 2016, is more complicated. This matter, at least initially, reflects the opposite situation of the court imposing a recommended treatment against the parents' consent. The Family Court of Western Australian made a series of determinations in regard to the treatment of a 6-year old boy subsequent to his diagnosis of a brain tumour – a medulloblastoma. The boy, Oshin Kiszko, underwent surgery, with the consent of his parents, in Perth, Western Australia, to resect the tumour at the end of 2015.

Consequently, his medical team advised a course of chemotherapy and radiotherapy to commence within 28 to 36 days following surgery to ensure his best chance of cure and survival. This treatment regime was acknowledged to be in accordance with accepted international practice. However, Oshin's parents objected to the proposed treatment regime on several grounds including a preference for natural therapies by the mother, an inability to care physically for Oshin due to a back condition from which the mother suffered, and the father's concern about taking further unpaid leave from work should Oshin be significantly impaired following the treatment.[24]

The ethics committee considered this matter on two occasions, and was divided in opinion. However, the committee concluded that it was "certain that Oshin's only chance of survival would be to embark on the standard treatment [...] and that the vast majority of parents faced with the same agonising choice would opt for a curative approach."[25] This was despite the expression by some on the committee that a more conservative approach involving palliative chemotherapy would be a valid and ethically sound alternative, suggesting that seeking orders to compel treatment against parental consent should be embarked upon

[23] Ibidem, para. 94.

[24] *Director of Clinical Services, Child & Adolescent Services*, FCWA 19 (n 17), paragraphs 26–29.

[25] Ibidem, para. 19.

with caution.[26] The committee condemned the use of alternative approaches to care, though, concluding that it was "not considered a rational approach" and that it was "ethically indefensible to impose such irrational beliefs on the lives of others."[27]

By mid-March 2016, investigations revealed that while Oshin's disease was progressing, the progress was "linear" rather than "explosive" and the medical team considered that they could not recommend palliation over treatment. Oshin's parents directly opposed the oncologist's recommended treatment regime. Their stance provoked the hospital into approaching the Family Court for an urgent order to intervene and provide the child with the best chance of a cure, that of a combined radiotherapy and chemotherapy regime. It was predicted that the recommended treatment would afford Oshin a 50-60% chance of survival after 5 years, the period at which the disease could be considered to be "cured."[28]

The Director, Clinical Services, Child and Adolescent Services of the Princess Margaret Hospital, brought this matter before the Family Law Court of Western Australia (WA). The Chief Justice, Thackray CJ, concluded that the evidence before him was that if Oshin received chemotherapy and radiotherapy, his chances of a long-term cure were good, despite the inevitable short to medium term negative impact of the treatment. This, together with his observation that most parents would choose intervention, led him to accede to chemotherapy commencing at the hospital. As Oshin's booking to receive radiotherapy was no longer available, the judge reserved judgement on the issue of radiotherapy until such time as chemotherapy had been administered. The judge appointed an Independent Children's Lawyer (ICL) to assist in the proceedings, funded by Legal Aid, WA.

The judge also noted that a holding order was in place prohibiting Oshin's removal from Australia and enforcing the withholding of the parents' passports.[29] This was on the background of a previous case before the Supreme Court of WA in which the parents were opposed to medical intervention and left the country with the child before the

---

[26] Ibidem, paragraphs 31–32.
[27] Ibidem, para. 38.
[28] Ibidem, para. 48.
[29] Ibidem, paragraphs 5-6.

court decision. The child received alternative therapy overseas and subsequently died soon after.[30]

In May 2016, when the Oshin case was reconvened before Thackray CJ, the situation had become even more heated. Two courses of chemotherapy had been administered with difficulties for clinical staff. The parents had erected a sign above the child's bed that read "Forced Chemo" during the first course and there were reports that Oshin's parents had sometimes refused to assist nursing staff in caring for and comforting Oshin.[31] His mother reportedly instead took a video recording documenting Oshin's distress.[32] The parents concurrently conducted a highly public campaign against the proposed treatment regime.

By May 2016, there had already been a considerable time delay in receiving full treatment, the boy's prognosis was diminishing, and a clinical dispute developed with respect to the recommended high-dose radiotherapy. This challenge emanated from an expert from NSW who had been approached by the ICL to provide an opinion. This paediatric oncologist, Professor Kellie, while vindicating Oshin's doctors' decision to reject palliative care measures in preference to curative treatment, raised the issue of the risk of long-term neurocognitive effects as well as other side-effects associated with the proposed high dose of craniospinal radiotherapy.

This report raised the possibility of a chemotherapy-only option and prompted a change in the approach by the parents, who now agreed to chemotherapy but not radiotherapy. However, Professor Kellie's recommended regime was to review MRI findings to gauge the child's response to chemotherapy with then a further discussion with the parents with respect to radiotherapy, at a reduced dosage.

In his deliberation, Thackray CJ identified "two social, moral and ethical questions."[33] The first of these was "whether greater emphasis should be placed on life itself rather than quality of life," and the second was "whether the first question should be answered by … the state or by the parents" who would otherwise make decisions in relation to the care of their child.[34] He thus raised the issue of the authority of the

[30] Ibidem, para. 60.
[31] Ibidem, FCWA 34, para. 13.
[32] Ibidem, FCWA 75, para. 69.
[33] Ibidem, FCWA 34, para. 65.
[34] Ibidem.

family in these decisions and commented that the court's power to countermand this authority should be used cautiously and sparingly. He concluded that:

> In the absence of a consensus of qualified medical opinion, there is, in my view, no role for the state in directing the parents to act in accordance with one entirely valid opinion in preference to another.[35]

He also raised other considerations in his determination, namely that there would now be a "substantial minority of parents" who would oppose radiotherapy at the doses proposed, and that:

> If Oshin were to have "forced" radiotherapy, … he would again be exposed to his parents' hostility and bitterness, potentially causing him even more psychological trauma.[36]

Thus, Thackray CJ concluded that enforced radiotherapy would not be in Oshin's best interests. However, in his judgment he acknowledged the time, effort, and energy of the treating doctors from Princess Margaret Hospital in attending court in order to preserve a "unique and valuable human life."[37] He added, "[i]t is therefore outrageous that they have apparently been denounced in social media for performing the very task that our society expects them to perform."[38]

The judge adjourned the hospital's application in regard to radiotherapy in light of this response from the parents to continue with further chemotherapy and ordered that they remain in Australia and receive treatment at the Princess Margaret Hospital for Children.

Oshin was reassessed by MRI scanning on July 7, 2016, following a further cycle of chemotherapy. There had been a significant improvement, again raising two difficult treatment options, both of which included chemotherapy and radiotherapy. When treatment was due to recommence at the end of July, Oshin's parents refused all treatment options other than palliation.

A further application to the Family Court was lodged to institute welfare orders in order to commence three cycles of consolidation

[35] Ibidem, para. 68.
[36] Ibidem, para. 71.
[37] Ibidem, para. 67.
[38] Ibidem.

chemotherapy, stem cell rescue and craniospinal radiation, or at least, consolidation chemotherapy with stem cell rescue. The hospital also sought injunctions against the parents allowing photographs or interviews, including any form of denigration of the staff caring for Oshin.

The hearing was conducted in August 2016 before O'Brien J, but by this time Oshin's chances of survival and cure had dropped considerably due to, in the words of the paediatric oncologist from Sydney, "the delays in delivering effective chemotherapy [that] have been unreasonable, avoidable and that have weakened or even negated any benefit that may have been achieved."[39] The counsel appearing for Oshin's parents conceded at the hearing that there was "an acknowledgement that the delays in treatment occurred because of decisions taken by the parents."[40]

The judge made many statements in his summation, including that:

> Considerable weight must be attached to the prolongation of life, but it is not absolute, nor necessarily decisive. Consideration must also be given to quality of life.[41]
>
> The multitude of elements which together comprise the best interests of the child are wide-ranging and multifaceted, and susceptible to a legitimate diversity of views. Best interests are values not fact.[42]
>
> The court's decision cannot be made in a vacuum. While the child whose best interests are being considered must be respected as a unique individual, each child lives in a family, and his welfare and best interests are inextricably interwoven with that family.[43]
>
> The court may legitimately be informed in its deliberations not only by the clinical opinions of the medical experts, but by their opinions as to the ethical issues. These opinions are born of specialist expertise, and experience in the value judgments inherent in the treatment of life-threatening ill-

[39] Ibidem, FCWA 75, para. 57.
[40] Ibidem, para. 68.
[41] Ibidem, para. 72.
[42] Ibidem.
[43] Ibidem.

> ness. By having regard to the views born of that expertise and experience the court does not abrogate its decision-making responsibility; rather it better inform the independent exercise of its decision.[44]

He also made it clear what the case was not about, namely: "parental rights, … the perceived power of the medical profession, … quality of life [being] more important than duration of life, (or) … the relative merits of traditional medicine and alternative or complementary therapies."[45]

Instead, he stated that he considered what was in Oshin's best interests at that point in time.[46] He held that Oshin was to undergo palliative care and any medical investigations and treatment that were necessary for this to take place as directed by the child's oncology team.[47] There were also instructions that the parents be restrained from denigrating on social network or speaking negatively about any hospital staff or their treatment plans to the media.[48] He ordered that Oshin was not to be interviewed, and for there to be no photographs or footage by any media outlet of Oshin in the vicinity of any hospital at which he attended.[49]

Oshin died in December 2016.

IV. Discussion

When one considers how Australia may have responded to the cases of Charlie and Alfie, it is clear that we have some not dissimilar case law involving the withdrawal or withholding of treatment upon which to base inferences.

Baby D, while representing a situation where there was accord between all parties, nevertheless was brought before the courts in part at least because there was uncertainty with respect to criminal charges that could be laid against clinical staff. The palliation that the baby may require would suppress brainstem function and hasten death, and nei-

[44] Ibidem, para. 73.
[45] Ibidem, para. 90.
[46] Ibidem, para. 91.
[47] Ibidem, para. 102.
[48] Ibidem.
[49] Ibidem.

ther the clinicians nor the ethics committee was prepared to proceed without legal sanction.

In Charlie's case, criminality was an issue as well in that the parents' lawyers threatened criminal proceedings against the hospital and its staff.[50] This brings into high relief that the process of end-of-life care is a difficult decision that inevitably, in both countries, necessitates, at times, a legal decision to protect clinical staff from criminal prosecution.

The case of Mohammed has distinct parallels, even down to the sharing of a mitochondrial disorder with Charlie. As with Charlie and Alfie, the judge, Garling P, hearing Mohammed's case, agreed with the opinion of medical staff, and mechanical ventilation was not provided for the child. Medical staff were not obliged to provide futile treatment, and due attention was focused on the pain and suffering of the child in an effort to determine the child's best interests.

It is difficult to assess the parents' reaction to this decision. As names were not released, there was no intense media attention directed to the family. The case has attracted notice within both medical and legal circles, however, due to its contribution to the principles at common law, especially with respect to professional obligations in futility of care matters.

Oshin's case on first view does not appear to be similar to either Charlie or Alfie given firstly that this case was brought because the parents refused treatment and secondly the ultimate decision by the judge to palliate as per the parents' wishes. It does nevertheless have some overarching issues that can be highlighted. The first issue is the additional problems associated with delays in conducting recommended clinical management during the period of protracted legal disputes. This time lag shifts the burdens and benefits equation. Oshin's prognosis declined over the series of hearings such that palliation ultimately became a less contentious option. Charlie's period of ventilation was lengthened during the period of disputation. The harm of extending intensive care had then already occurred, so the question arose whether the additional period of intensive care for a limited trial of treatment

---

[50] *Great Ormond Street Hospital v Yates, Gard & Gard* [2017] EWHC 1909, para. 2.

did in fact signify an unwarranted extension of further pain and suffering.[51]

Also an important element in common between the matters is that the cases necessitated a decision when there was some level of disagreement between the medical evidence. As previously cited, Thackray CJ presiding over the second Oshin hearing stated, "in the absence of a consensus of qualified medical opinion, there is … no role for the state in directing the parents to act in accordance with one entirely valid opinion in preference to another."[52]

While this seems to imply that Australia may have allowed the transfer of a child to the US for experimental treatment had this been the set of circumstances before the court, there are two considerations that indicate that this may not have occurred. The first is the order of Thackray CJ to uphold the previous orders of the magistrate to prohibit Oshin from being removed from Australia.[53] Oshin should, in his opinion, remain in Australia to receive evidence-based care, an opinion mirrored by O'Brien J presiding over the third hearing. He directed Oshin's parents to "meaningfully engage with the oncology team at Princess Margaret Hospital and with any other health professionals as recommended by the treating oncologist."[54] The parents were thus directed to remain with the same clinical team so that the child could be monitored and palliated under specialist attention.

The second consideration in comparing the outcome of the Oshin matter with that of Charlie or Alfie is that the degree and type of medical inconsistency is not necessarily equivalent. In Alfie's case, there was overwhelming medical opinion on the degree of brain deterioration. Progressive MRI scanning demonstrated relentless neurodegeneration. With respect to the issue of Alfie's cognition of pain, the specialist paediatricians considered that while it was unlikely, it could not be determined conclusively that he was not aware of pain and discomfort. And

---

[51] D. Wilkinson, J. Savulescu, "Hard Lessons: Learning from Charlie Gard Case," *Practical Ethics: Ethics in the News* (blog published by the University of Oxford, July 24, 2017). http://blog.practicalethics.ox.ac.uk/2017/07/hard-lessons-learning-from-the-charlie-gard-case/ (accessed November 17, 2018).

[52] *Director of Clinical Services, Child & Adolescent Services*, FCWA 34 (n 17), para. 68.

[53] Ibidem, FCWA 19, para. 83.

[54] Ibidem, FCWA 75, para. 102.

Dr R, consultant in paediatric neurology, added a disquieting comment. He explained that also at issue was the possibility that the child may feel pain without any benefit of a sensation of comfort from the voice or touch of his parents.[55] There was also agreement with respect to life-support measures being able to sustain Alfie, but medical disagreement with respect to withdrawal of therapy and whether, should the child be transferred to either Italy or Germany, this signified continued suffering, or, worse, increased suffering.

Similarly, medical ethicist Dominic Wilkinson quotes the barrister for Great Ormond Street Hospital (GOSH) as saying that the hospital had been unable to identify a single UK intensive care unit that was prepared to sustain treatment for Charlie.[56] And Mr Justice Francis in his judgement makes reference to the expert independently instructed by Charlie's parents, who agreed with the experts from GOSH.[57] The implication of this is that the vast consensus amongst qualified UK medical specialists concurred with ending life-support despite the offer of the controversial experimental treatment, nucleoside bypass therapy, for Charlie's very rare form of mitochondrial disease at his stage of brain damage.

In contrast, in Oshin's case, the independent oncologist, while cautious about radiotherapy, still wanted to proceed with more intensive chemotherapy, especially given the boy's encouraging response on serial MRI scanning. Ultimately the delays in providing comprehensive possibly curative treatment meant that the goalposts shifted, altering prognosis in the process. The judge, while ultimately agreeing with palliative measures, made a point of saying that he was aware that both the staff caring for Oshin and the independent oncologist recommended curative treatment and that neither had ever before ceased the treatment of

---

55 See *Alder Hey Children's NHS Foundation Trust v Evans, James & Evans* [2018] EWHC 308.

56 D. Wilkinson, "The Sad Case of Charlie Gard and the Rights *and Wrongs* of Experimental Treatment," *Practical Ethics: Ethics in the News* (blog published by the University of Oxford, July 24, 2017). http://blog.practicalethics.ox.ac.uk/2017/07/the-sad-case-of-charlie-gard-and-the-rights-and-wrongs-of-experimental-treatment/ (accessed November 17, 2018).

57 *Great Ormond Street Hospital* (n 50), para. 3.

a child who was responding positively.[58] His judgement is consistent with the observations made by academic Joanna Manning who, on researching like decisions in New Zealand, commented that:

> there are no decisions authorising treatment overriding parental refusals where the prognosis is less than a 50 per cent chance of survival, or where the condition is merely treatable as opposed to curable, and the court is asked to balance quantity against quality of life.[59]

The judges then over these three matters confronted different degrees of medical divergence, especially in regard to prognosis and the benefit of any proposed intervention. What is significant, however, is their consistency in assessing this divergence in court, calling and testing all medical evidence.

But possibly the most striking similarity is the part that media outlets and social networking have played in this Australian case and the UK cases. Oshin's plight, like that of Charlie and Alfie, was a very public event, with the series of hearings using his name and location openly throughout.[60]

---

[58] *Director of Clinical Services, Child & Adolescent Services*, FCWA 75 (n 17), para. 94.

[59] J. Manning, "Parental Refusal of Life-prolonging Medical Treatment for Children: A Report from New Zealand," *Journal of Law and Medicine* 8.1 (February 2001): 275.

[60] The many cases that have been heard previously across Australia in which the best interests of the child have been in dispute are, as a general rule, heard in a closed court. The court takes great pains to protect the anonymity of all concerned, withholding the names and location of the parents, and at times even the names of the representing counsels and solicitors, and the hospitals and treating clinicians. See C. Smith, "Parens Patriae Jurisdiction." In a case heard before the Supreme Court, Brisbane in 2015, the press accessed the file due to the inadvertent filing of documents without due attention to the legislative non-publication provision. See, *The Hospital v T and Anor* [2015] QSCA 185. As commented by Douglas J, this led to "distressing invasions of the privacy both of the respondents and the doctors…treating the child" at para. 9. The judge, too, in Oshin's case, attempted to staunch the media attention and public denigration of staff by issuing orders to this effect, a step that is in keeping with previous efforts to manage these matters with sensitivity. See *Director of Clinical Services, Child & Adolescent Services* FCWA 75 (n 17), para. 6.

All three sets of parents have in effect encouraged public participation in their cases, such that the public has often opted for the parents' side of the story with protests against clinical staff. The parents have taken measures that have included online campaigns, crowdfunding, public relations consultation, and petitions to other bodies, which range from governments through to religious leaders.

And the public has responded avidly to the widely reported campaigns, despite, by and large, having a limited grasp of the clinical and ethical issues confronting both the medical and legal teams in protecting the best interests of the child. This crowd obviously can be twisted into either wanting treatment or withdrawing treatment over the very same issues of prognostic chances and degrees of suffering. And ideological movements that range from pro-life groups to anti-medical campaigners grasp the opportunity to promote their causes in the media.[61]

One can only feel profound sympathy for the predicament of these parents, and trust that they believe that they were pursuing the correct course for their children. But one must also feel sympathy for the clinical staff who have been harangued and vilified with little opportunity to countermand the hostile allegations while they try to offer these very unwell children and their emotionally traumatised parents as much comfort and clinical care as they can.

Australian QC Ian Freckleton asks whether there should then be some primacy given to the views of the parents.[62] He comments that Oshin's case has produced a most unfortunate precedent in that it gives a signal to others that the manipulation of "the crowd" through the media ensures the outcome that the parents are demanding. Oshin, he argues, was essentially used as a weapon, and he cautions that, as a result, Oshin's voice was not heard.

While there is some validity to this observation, the courts, both in Australia and the UK, took pains to provide an independent "voice"

---

[61] G. Hinsliff, "Alfie Evans' Parents Needed Help. The Vultures Came Instead," *The Guardian* (published on April 27, 2018). https://www.theguardian.com/commentisfree/2018/apr/26/alfie-evans-parents-activists (accessed July 23, 2018).

[62] I. Freckelton, "Parents' Opposition to Potentially Life-saving Treatment for Minors: Learning from the Oshin Kiszko Litigation," *Journal of Law and Medicine* 24:1 (2016).

for the child. Thackray CJ appointed an Independent Children's Lawyer (ICL) to represent the best interests of Oshin, and it was the ICL who approached Professor Kellie to provide an independent medical opinion. Baby D similarly had an ICL appointed through Legal Aid. The aim of the ICL is to "remain independent, objective and focused upon promoting the child's best interests in all dealings throughout the proceedings."[63] And, consistently, Charlie and Alfie had Guardians appointed to independently represent the children's interests throughout the legal process in the UK.

This observation naturally leads to the final point of resemblance, which is the process of both medical and legal assessments used in these matters, and the uniformity of the principles that have been adopted. Australia and the UK share a strong public health system wherein clinicians certainly do not set out to withdraw life-sustaining treatment, but nor do they advise on pursuing treatment options that are ultimately futile and painful.

It is apparent that in both jurisdictions, treatment plans are developed with parental consultation over lengthy periods of discussion. During disputes, additional measures are used such as ethics committees and mediations in an effort to reach a consensus about care. Medical decisions are made with caution when there is either dispute, such as with the matters of Oshin, Charlie and Alfie, or when the proposed procedure or treatment regime ventures towards areas that may surpass the zones of parental consent. "Marion" and the request of her parents for sterilisation, and Baby D with the likelihood of suppressing brainstem function are examples of the latter.

When the situation becomes irreconcilable, the judiciary, under the jurisdiction of *parens patriae*, becomes the final arbiter. The judges in their determinations have applied the test of whether continued treatment or withdrawal of treatment is in the best interests of the child. As Thackray CJ, as previously quoted, observes, the judge in these circumstances is confronted with two essential moral and ethical questions of "whether greater emphasis should be placed on life itself rather than

---

[63] Guidelines for Independent Children's Lawyers (2013), para. 6.4. https://www.legalaid.nsw.gov.au/__data/assets/pdf_file/0012/20046/ICL-Guidelines.pdf (accessed November 20, 2018).

quality of life" and whether this question needs to be answered by the state or the parents.[64]

Given the extreme media attention that has initiated a wide commentary on the case of Charlie in particular, this issue of quality of life versus life itself has generated alternate medical ethical opinions. Peter Singer and Julian Savulescu, both Australian ethicists, have written in support of Charlie receiving experimental treatment on the grounds of a "life worth living."[65] But this therapy was advocated despite having not been previously tried for this rare mutation, not even on mice, prior to a clinical assessment being conducted by Professor Hirano from the US,[66] and in the presence of irreversible brain damage.[67] As commented by Alistair Stewart in response to the contributions to the debate offered by Savulescu and Singer, it is "the space between probability and absolute certainty into which traders of hope can insert themselves."[68]

Wilkinson (another Australian ethicist) observes on this issue that decisions with respect to experimental treatment involve a number of steps that include seeking experts with relevant expertise, obtaining additional opinions, and then approaching an impartial arbiter.[69] He argues that these are steps taken by Mr Justice Francis in the Family Division of the High Court and that his decisions were upheld in the Appeals Court, the Supreme Court and in the European Court of Human Rights.

There has certainly been a general observance of this process by the recognised impartial arbiters, the judges in both jurisdictions, when testing medical evidence throughout this case law. The judges have

---

[64] *Director of Clinical Services, Child & Adolescent Services*, FCWA 34 (n 17), para. 65.

[65] J. Savulescu, P. Singer, "Unpicking What We Mean by Best Interests in Light of Charlie Gard," *The BMJ Opinion* (blog uploaded August 2, 2017). https://blogs.bmj.com/bmj/2017/08/02/unpicking-what-we-mean-by-best-interests-in-light-of-charlie-gard/ (accessed November 24, 2018).

[66] *Great Ormond Street* Hospital (n 50), para. 9.

[67] Ibidem, para. 15.

[68] A. Stewart's blog comment on "Unpicking What We Mean by Best Interests in Light of Charlie Gard," *The BMJ Opinion* (blog uploaded August 2, 2017). https://blogs.bmj.com/bmj/2017/08/02/unpicking-what-we-mean-by-best-interests-in-light-of-charlie-gard/ (accessed November 24, 2018).

[69] D. Wilkinson, "The Sad Case."

sought expert advice on futility of care, likely prognostic outcomes and most importantly, levels of cognition and suffering. But too, they have listened with sensitivity to parental concerns and opinions and, Oshin being a ready example, have taken into consideration the family and social network of the child.

On the second issue of state versus parental determination, both Australia and the UK come back to the fundamental issue of protecting the child and see this as a function of the judiciary. Public commentary, particularly in response to Charlie and Alfie, has been plentiful on this aspect, but, as commented by UK criminal barrister Matthew Scott, "well-intentioned calls to give parents more legal rights over these terrible end of life decisions are fundamentally mistaken."[70] He applauds the UK's "courageous and independent family court judges who follow clear and long-established principles of law."[71]

These "principles of law" use past case law as guidance in both Australia and the UK, but the overwhelming ethos is that each case needs to be considered in the interests of the individual child in the context of a legal responsibility to society to protect the life and wellbeing of the young and vulnerable. It is significant that these cases are heard not from the perspective of wasting public money on futile care, or the inappropriate use of limited resources, but on the consistent fundamental concern of what is in the best interests of this particular child at this time.

V. Conclusion

Would Australia have responded differently to Charlie and Alfie? An overview of these four Australian cases has illustrated a consistency of both medical approaches and legal principles with those of the UK cases of Charlie and Alfie. In particular, it is significant that these cases adopt a comparative process of assuring that the child is represented separately from other parties; of seeking medical opinion and testing the opinion with independent advice; addressing the particularities of the medical and social situation of each child that comes before the

---

[70] M. Scott, "The Tragic Case of Alfie Evans," *Quillette* (published on April 28, 2018), https://quillette.com/2018/04/28/tragic-case-alfie-evans/ (accessed November 20, 2018).

[71] Ibidem.

court; and giving the parents a voice. Given this degree of consistency, it is quite possible that the outcomes would have been the same.

Further, from this comparison, it is clear that Australia is grappling with the parallel issues and difficulties associated with these complicated and highly emotional end-of-life choices in the face of new factors. These include the increasing participation of the public with crowd-funding and social media interests; the "expert" parent who accesses the internet for possible treatments and cures; the increasing role of medical ethics in resolving these matters; and the continued advances in medicine and life support that pose further challenges to the already complex and multifaceted decisions around medical intervention and the withholding of treatment.

Chapter 11

Medical Futility in Czech Paediatrics: At the Edge of Law, Bioethics, and Medicine

Helena Krejčíková
Charles University in Prague, The Czech Republic

I. Introduction

End of life still represents one of the greatest taboos in the Czech Republic and, until recently, was hardly ever publicly discussed. Medicine has been presented as victorious, with increasingly advanced treatment options offered to patients, so the terminal stage of an incurable illness is often interpreted as a failure of either the doctor, or medicine, generally.[1] Doctors are not trained otherwise; they often feel the (if not legal, then surely ethical) obligation to save lives by heroic measures[2] or at any costs. They have no systematic training in communication on fatal prognoses, nor is there the time or space in daily clinical praxis for a proper discussion regarding end-of-life decision making. In addition, because of legal concerns, the whole system tends towards defensive medicine, which is not provided for the benefit of the patient concerned, but for the legal protection of the doctor.

However, the problem of medical futility cannot be solved by denying that, at some stage, curative medicine has no option to offer. Pretending otherwise gives no benefit to the patient, rather the contrary.

---

[1] As stated, among others, by Prof. Marta Munzarová in the interview on the role of bioethics. In: D. Povolný, *Munzarová: Lékařská Etika je Nepostradatelnou Součástí Medicíny* [Munzarová: Bioethics is an Indispensable Part of Medicine] (Brno: Masarykova Univerzita, 2009).
https://www.online.muni.cz/veda-a-vyzkum/1675-munzarova-lekarska-etika-je-nepostradatelnou-soucasti-mediciny (accessed November 29, 2018).

[2] The duty to save life regardless of the doctor's own safety was stated by law until 2012.

If there is no clear information on medical futility, all further end-of-life decision-making suffers from a lack of transparency and reviewability, as the premise itself cannot be challenged. Even in cases where medical futility is appropriate, not admitting it might harm the patient either by not providing the adequate treatment (typically, early palliative care), or by not providing the patient with adequate information and the time to resolve his/her personal matters, etc. Therefore, accepting medical futility as an inevitable part of providing healthcare is crucial in order to provide it properly.[3]

As this chapter shows, the end of life of a minor still remains an area which is not regulated unambiguously by law. Czech medical practice has no reliable legal framework for dealing with incurably ill children, not to mention difficult cases like those of Charlie Gard or Alfie Evans. With two already-published guidelines covering futile care for adults and one preparatory material targeting futile care in paediatrics, end-of-life decision-making might – at least from the medical point of view – become less intuitive and more balanced. The parameters of legal proceedings regarding *ad hoc* medical futility, on the other hand, are still to be clarified. Besides introducing the relevant legal regulations and guidelines regarding medical futility, this chapter will also discuss the most important aspects of such proceedings as can be assumed considering the relevant court decisions and current prevailing legal opinion.

II. Providing healthcare to a minor under Czech Law

During the last two decades, Czech Medical Law has experienced a significant shift from the post-Soviet paternalistic doctor-patient approach towards greater respect for the patient's autonomy. Due to the ratification of the Convention on Human Rights and Biomedicine in

[3] Similar reasoning is found in the Guidelines on Futile Care regarding adults, i.e. the consensual opinion of the Czech Medical Society of J. E. Purkyně on providing palliative care to patients with irreversible organ failure and the Recommendation of the Board of Directors of the Czech Medical Chamber No. 1/2010 on decision-making regarding the conversion from intensive care to palliative care for terminally ill patients unable to express their wishes. Available (in Czech only) at:
https://www.urgmed.cz/postupy/cizi/2009_paliativni.pdf and
https://www.lkcr.cz/doc/cms_library/dop-1-2010-31.pdf
(accessed November 29, 2018).

1997,[4] the passing of the Act on Healthcare Services in 2011,[5] and the recodification of the Civil Law via the Civil Code of 2012,[6] the position of a patient as an equal partner to a doctor has been progressively strengthened, although still with substantial limitations.[7] However, the situation differs slightly when healthcare is provided to a minor.

The legal position of a minor patient, i.e. a patient younger than 18 years, who has not been granted full legal capacity by a court,[8] is mainly governed by section 35 of the Act on Healthcare Services, together with section 95 and sections 100 and 101 of the Civil Code. This legislative solution could partly be understood as the unwanted consequence of a lack of coordination between the authors of the two Acts (namely the Ministry of Health and the Ministry of Justice), as the legislative procedure of the latter codex did not reflect the existence of the first one at all.[9] The result causes unsatisfactory legal uncertainty, which is only slowly being overcome by legal doctrine and case law.

---

[4] The Czech Republic signed the Convention on Human Rights and Biomedicine in 1998 and ratified it in 2001. The Convention is available online at https://rm.coe.int/168007cf98 (accessed November 29, 2018).

[5] Act No. 372/2011 Collection. Available (in Czech only) online at https://www.zakonyprolidi.cz/cs/2011-372 (accessed November 29, 2018).

[6] Act No. 89/2012 Collection. Available online at: http://obcanskyzakonik.justice.cz/images/pdf/Civil-Code.pdf (accessed November 29, 2018).

[7] Thus, e.g. the patient is not entitled to request the withdrawal of mechanical life support via his/her previously expressed wish. For more detail, see H. Peterková, "Previously Expressed Wishes in the Czech Republic - A Missed Chance of the Czech Legislator," *Medicine and Law* 32:4 (2013): 429-440.

[8] According to section 37 paragraph 1 of the Civil Code, if a minor without full legal capacity applies to a court to be awarded legal capacity, the court shall grant the application if the minor has reached the age of sixteen years, if his ability to provide for his maintenance and take care of his welfare has been proven, and if the legal representative of the minor consents to such an application. In other cases, the court shall grant the application if it is in the interests of the minor for serious reasons.

[9] As can be seen from the Explanatory Report to the Civil Code, referring to a previous legal regulation (from 1966). The Report is available in Czech only at: https://apps.odok.cz/attachment/-/down/RACK7MFCFLKI (accessed November 29, 2018).

Unless there are reasons for providing healthcare without any informed consent (such as a state of necessity in which healthcare is needed to save a patient's life or health, or in the case where there is suspicion of maltreatment by parents, etc.), informed consent is required in order to provide healthcare to a minor. Under the Act on Healthcare Services, if it is adequate to the intellectual and volitional maturity of the minor patient's age, the patient himself/herself can give his/her informed consent to the provided healthcare services.[10] Nevertheless, under the Civil Code, this rule applies only in usual matters and if it is an intervention not resulting in any permanent or serious consequences.[11]

Under other circumstances, healthcare shall be provided with the consent of the minor's legal guardian (mostly parents in a joint decision[12]). If the minor is not competent to give his/her informed consent, but has already reached the age of 14 years, any dispute between him/her and his/her parent/s regarding a treatment shall be solved by a court.[13] In addition, if the minor is fully incompetent, and a highly risky medical intervention or a medical intervention with severe side effects is taken into consideration, the court shall approve it before it is provided, unless the intervention is needed urgently.[14] Especially the last rule regarding the duty to notify a court, introduced by section 101 of the Civil Code which was adopted from the German Civil Code,[15]

[10] Section 35 paragraph 1 of the Act on Healthcare Services.

[11] Section 95 of the Civil Code.

[12] Cf. Section 692 paragraph 1 of the Civil Code: Family matters, including the choice of the place of the family household or, where applicable, the household of one of the spouses and other family members, in particular children who have not yet acquired full legal capacity, and the way of life of the family, are to be agreed by spouses.

[13] Section 100 of the Civil Code.

[14] As stated in section 101 of the Civil Code verbatim, if the integrity of an individual incapable of judgment is to be interfered with in a way resulting in permanent, irreversible, and serious consequences or a way associated with a serious threat to his life or health, the intervention may only be carried out with the leave of a court. The wording of the provision causes some doubts whether the court shall also be notified of the withholding or withdrawal of the treatment.

[15] Bürgerliches Gesetzbuch. Section 1904 of the German Civil Code regulates what is known as approval of the custodianship court in case of medical treat-

might be considered crucial for clinical practice; it is nonetheless hardly ever followed, as to be discussed below.

In order to understand the factual position of a minor and his/her parents in a healthcare system, it needs to be emphasised that what is constructed in theory as an informed choice of treatment (out of many alternatives, always with one alternative consisting of a refusal of all possible treatments), is often limited to an informed consent for a treatment suggested by a doctor. Czech Medical Law of course offers the possibility of refusing a treatment indicated by a doctor.[16]

However, where a minor is concerned, the doctor is often allowed by law to pursue the treatment not only without any informed consent, but also against the will of the entitled person(s)[17] in any case (if urgent or acute care is needed in order to save the life or health of the patient, or to prevent sudden or intense pain, or a sudden state in which the patient can endanger himself/herself or his/her surroundings).[18] Furthermore, if the doctor comes to a conclusion that, by refusing treatment, the parents are actually endangering the patient, he/she[19] can set up a legal procedure which can end with a court ordering the treatment.[20]

On the other hand, if the doctor is not willing to provide a patient with a certain treatment on the basis of his/her clinical judgement, the

ment (Genehmigung des Betreuungsgerichts bei aïrztlichen Maßnahmen). https://www.gesetze-im-internet.de/englisch_bgb/englisch_bgb.pdf. (accessed November 29, 2018).

[16] Section 34 paragraph 4 of the Act on Healthcare Services regarding an adult. However, concerning a minor, the law regulates no explicit right to refuse treatment. According to section 35 paragraph 1 of the Act on Healthcare Services, the disagreement of the patient shall be taken into account in accordance with the degree of achieved maturity.

[17] I.e. the patient and/or the parents or other legal representatives of the child (most likely: the guardian).

[18] Section 35 paragraph 4 together with section 38 paragraph 4-5 of the Act on Healthcare Services.

[19] More precisely, the healthcare provider, i.e. often the hospital.

[20] Section 6 of the Act No. 359/1999 Collection, on the Sociolegal Protection of Children. Available online (in Czech only) at:
https://www.zakonyprolidi.cz/cs/1999-359 (accessed November 29, 2018).

patient or his/her parents have limited options to request it.[21] The mutual relationship between a doctor/provider of healthcare services and the patient of course differs from the relationship between a client and provider of other services, because of the very delicate nature of the services concerned; with the patient (or his/her parents) being obviously the weaker party to the contract.[22]

The whole legal construction has one perfectly justifiable goal in mind, which is to ensure the maximum protection of the life and health of a minor and his/her proper access to *adequate* healthcare. As the Czech Constitutional Court ruled in 2003, the life and health of a child is considered the highest value in society, above all others, which deserves ultimate protection.[23]

While stressing the paramount duty of not treating a patient inadequately, Czech medical practice has not made enough effort to ensure that the patient will at the same time not be treated excessively. Indeed, what could be regarded as a greatly protective approach seems to be predominant and causes actual problems if the life and health of a minor can possibly not be saved due to the patient suffering from an incurable disease in its terminal stage. Although there is no legal right of a patient or his/her parents to order the doctor to provide a particular treatment, because of defensive reasons, the doctor often follows their wish and treats the patient intensively, regardless of medical judgement. This is far too common in terminally ill minor patients, where healthcare workers hesitate to let a child die without, for instance, at-

[21] Consisting of e.g. changing the doctor, as there is the right to free choice of a healthcare provider under section 28 paragraph 3(c) of the Act on Healthcare Services.

[22] Although healthcare in the Czech Republic is publicly funded, the doctor-patient relationship is governed by contract. Cf. among others, the Explanatory Report to the Civil Code, regarding sections 2636 and 2637, available online (in Czech only) at:
http://obcanskyzakonik.justice.cz/images/pdf/Duvodova-zprava-NOZ-konsolidovana-verze.pdf (accessed March 13, 2019).

[23] The decision of the Czech Constitutional Court, No. III.ÚS 459/03. Available online (in Czech only) at:
http://nalus.usoud.cz/Search/GetText.aspx?sz=3-459-03
(accessed November 29, 2018).

tempting resuscitation.[24] Even though court involvement via section 101 of the Civil Code should have brought more clarity to decision-making in end of life cases of incompetent patients, it is still not the case, as clinical practice is unwilling to report such cases.

III. Medical futility

The concept of medical futility has not been explicitly recognised by the Czech legislator. However, what is in general futile treatment can be deduced from the definition of a proper care standard (in statutory wording: healthcare service provided on an appropriate level) under section 4 paragraph 5 of the Act on Healthcare Service, i.e. healthcare provided according to science and acknowledged medical guidelines, with respect to the individuality of the patient, with regard to existing conditions and objective possibilities. Although the definition might suggest otherwise, medical guidelines do not present a binding legal instrument in the Czech Republic and it is always the court which decides whether *ad hoc* treatment was provided on an appropriate level (*lege artis*) – among others, on the basis of expert opinion. Nevertheless, if there are medical guidelines covering the medical situation concerned, the expert cannot easily ignore it – e.g. should the expert come to the conclusion that the treatment was not properly provided, even though the doctor followed the guidelines, the expert must explain his opinion convincingly.

With regard to adult patients, there are two available documents dealing with medical futility, i.e. the Consensual Opinion of the Czech

---

24 J. Matějek, "Limity o Těžce Postižené a Těžce Nemocné Děti" [Limits of Care of Seriously Disabled and Ill Children], *Pediatrie pro praxi* 13:1 (2012): 56. Also available online (in Czech only) at: http://www.pediatriepropraxi.cz/artkey/ped-201201 (accessed November 29, 2018). The same conclusion can be brought from the survey conducted by Bc. Jana Bubláková in Jihlava Hospital. In J. Bublaková, *DNR Jako Eticko-ošetřovatelský Problém* [DNR as an Ethico-Nursing Problem], Master Thesis. (České Budějovice: Jihočeská Univerzita v Českých Budějovicích, Zdravotně Sociální Fakulta, 2012), 57. https://theses.cz/id/4qtuov/DNR_jako_eticko-osetrovatelsky_problem2.pdf (accessed November 29, 2018).

Medical Society of J. E. Purkyně[25] on providing palliative care to patients with irreversible organ failure, and the Recommendation of the Board of Directors of the Czech Medical Chamber No. 1/2010 on decision-making regarding the conversion from intensive care to palliative care for terminally ill patients unable to express their will (further also referred to as Adult Guidelines). Despite the recognition from professionals, these documents are not legally binding,[26] even though their factual impact might be considerable, as further discussed.

Both documents (which in fact are identical because the Czech Medical Chamber took over the Consensual Opinion in order to acknowledge it to some extent)[27] define medical futility in the context of the terminal stage of an incurable disease. According to the Adult Guidelines, "futile and purposeless treatment" does not save life, preserve health, or maintain the patient's quality of life; furthermore, it is not in the interest of the patient as it has no potential to aid him/her and burdens the patient because of unnecessary suffering or risk of complications. Such a treatment shall not be provided.

In 2010, when the Adult Guidelines were introduced, they were interpreted by the public as a highly controversial effort to legalise euthanasia.[28] Regardless of its initial reception, the contribution of the Adult Guidelines to clinical practice remains unchallenged. For the first time,

---

[25] The Czech Medical Society of J. E. Purkyně is a voluntary and independent society, consisting of healthcare workers and composed of many societies on the basis of medical specialisations.

[26] From a legal point of view, the Czech Medical Society of J. E. Purkyně, is (merely) a civil society with no authority to introduce any legally binding regulations to the public, still less for their members. This should not be confused with the Czech Medical Chamber founded by Act No. 220/1991 Collection, as membership in the latter is a mandatory prerequisite for being employed as a doctor in the Czech Republic. If the Czech Medical Chamber had wanted to issue guidelines (opinions) binding on all doctors, they could have done so under section 2 paragraph 2 (i) of the Act.

[27] Although not in the form of a binding opinion (see above).

[28] e.g. The Hyde Park programme, broadcast by Czech Television on 2 March 2010, in the episode titled "Eutanazie po česku?" [Euthanasia the Czech Way?] and the comment published on 11th March 2010 on the website http://www.zdn.cz/denni-zpravy/komentare/polovicata-eutanazie-450239 (no longer available) under the title "Polovičatá Eutanazie" [Halfway Euthanasia].

it was publicly admitted that at some point medicine can exhaust its curative possibilities. Furthermore, the Adult Guidelines relativise the absolute protection of life, as they state that a patient's life shall not be maintained irrespective of its quality or the dignity of the person concerned. This view is fully consistent with the principle of proportionality which enables the two colliding fundamental rights to be balanced.[29]

Despite the generally applicable definition of "futile and purposeless treatment" stated in the Guidelines, clinical practitioners felt there was a lack of a specific document taking into account the delicate aspects of paediatric care. Its conception became one of the main goals of the recently established Paediatric Section of the Czech Society for Palliative Medicine,[30] with the perspective that it might preferably be joined by other specialist societies of the Czech Medical Society of J.E. Purkyně.

A. Futile care in paediatrics

In January 2018, the Paediatric Section finalised a document called "Preparatory Material for Consensual Opinion on Providing Life-Sustaining Treatment"[31] (further also referred to as the "Paediatric Guidelines"). This is a paediatric version of the Adult Guidelines on Futile Care. The title "Paediatric Guidelines" was deliberately chosen in order to minimise any *a priori* negative connotation of the words "withholding," "withdrawal," etc. In addition, the title of the document might suggest an omnipresent positive and proactive approach towards the patient. Instead of focusing exclusively on which particular treatment is or is not futile and should not be provided, the document highlights the fact that, under all circumstances, there is still an adequate

[29] The decision of the Czech Constitutional Court, No. PL ÚS 4/94. Available online (in Czech only):
https://nalus.usoud.cz/Search/GetText.aspx?sz=Pl-4-94 (accessed November 29, 2018).

[30] The working group for Paediatrics was established in November 2015 as part of the Czech Society for Palliative Medicine and was transformed into a Section two years later.

[31] The Guidelines have not been officially published yet, as they are now subjected to more discussion by other specialist societies.

treatment available to the patient, which is palliative care to alleviate distress and suffering.

Unlike the guidelines regarding adults, the Paediatric Guidelines do not use the term "futile treatment." This is explained by the fact that medical futility can only rarely be precisely defined, unless it is an obvious physiological futility.[32] More often, it should be decided what extent of diagnostics and/or medical care is purposeful in order to achieve the outlined goals of the treatment, and thus the term "purposeless treatment" is used in the Paediatric Guidelines.

The document is presented as procedural guidelines for decision-making in a situation in which either the patient (or the parents) or the doctors or all of them together, consider a treatment as purposeless, and therefore contraindicated. Such decision-making can be complicated by the fact that a minor is not entitled (not even an emancipated one[33]) to formulate what is known as a "previously expressed wish" (advance directives, living will) under Czech Law, nor do parents have the right to make such a decision on behalf of the patient.

Additionally, the guidelines also present a procedural solution for a situation in which there is a dispute between the patient (his/her parents) and doctors on whether the treatment is purposeless or not. In some particular situations, the guidelines anticipate the court's involvement, as further shown.

As the Paediatric Guidelines clarify, decisions regarding life-sustaining treatment must in general not only be subjected to law, but also to ethical principles (autonomy, beneficence, non-maleficence, and justice). Specifically, some other principles to be followed are as below:

---

32 O. Sláma, "Lze Objektivně Vymezit, Která Léčba je u Nevyléčitelně Nemocného Pacienta „marná"? Několik poznámek z perspektivy onkologické paliativní péče." [Is it Possible to Define Objectively which Treatment regarding Terminally Ill Patients is Futile? Several Remarks from the Oncologico-palliative Care Perspective], in *Smrt a umírání* [Death and Dying], ed. David Černý and Adam Doležal (Prague: Ústav státu a práva AV ČR, 2013), 216-223 (in Czech only).

33 Under section 30 paragraph 2 together with section 37 of the Civil Code, before reaching the age of majority, full legal capacity is acquired by being granted legal capacity by the court (known as emancipation) or by entering into a marriage.

1. Withdrawal or withholding of life-sustaining treatment must not be confused with euthanasia (life-termination upon request).
2. Adequate treatment of pain and discomfort must never be labelled as "purposeless treatment" under the Paediatric Guidelines. The ethical principle of double effect is acknowledged, as it is proclaimed to be ethically admissible to provide the patient with treatment maximising the quality of his/her life, even though it might shorten life.
3. The wish of the patient[34] who is able to express it should be taken into account to the maximum extent allowed by law.
4. If a treatment is considered purposeless, the patient is still provided with medical care. It is always ensured that the patient is provided with care of an adequate extent and intensity. Unlike artificial nutrition and hydration which might be – if purposeless – withdrawn or withheld,[35] the patient is always provided with oral nutrition and hydration if he/she is capable of consuming them.
5. A decision that life-sustaining treatment is purposeless is based on the evaluation of the expected benefits of the treatment (life prolongation, symptom alleviation) in proportion to the burden which the treatment represents for the patient (side-effects, invasiveness, etc.). The crucial element is the consideration of the estimated quality of the further life of the patient and the extent to which the patient will be able to conduct normal daily activities and be involved in normal life situations.[36]
6. Decision-making regarding life-sustaining treatment is always to some extent subjected to a prognostic uncertainty; the doc-

[34] Given the fact that the upper age range for the Paediatric Guidelines is 18 years, some of the patients concerned would be able to express their wishes.

[35] If the patients concerned are not able to orally consume nutrition and fluids, in those cases in which they are considered medically purposeless, nutrition and hydration will not be provided artificially via e.g nasogastric tube.

[36] As assessed through the International Classification of Functioning, Disability and Health (ICF). WHO. ISBN 92 4 154542 9. The Classification is available online at: http://apps.who.int/classifications/icfbrowser/ (accessed November 29, 2018).

tor who makes such a decision has a duty to carefully consider the risks, alternatives and benefits of the treatment, to write down his/her deliberation in the medical records of the patient, and to communicate it to the persons legally entitled to it.
7. From an ethical point of view, withdrawal of treatment is equivalent to withholding of treatment.
8. The decision makers must never confuse the values and interests of the patient with their own.
9. The costs of a treatment are irrelevant for decision-making regarding the purposelessness of the treatment. However, this stipulation does not rule out the possibility that the treatment will not be provided because of its costs.
10. Every patient is unique – a decision regarding the purposelessness of a treatment must be made after complex discussion. The decision must not be based merely on particular elements, such as diagnosis, clinical state, prognosis, ethical and socio-economic conditions, etc.

B. Decision-making regarding purposeless treatment

The primary importance of the Paediatric Guidelines is the explicit acknowledgement that under specific conditions, not even a minor should be provided with all (intensive and/or invasive) care available to prolong his/her life regardless of its quality. Of secondary importance is the introduction of particular procedural steps of the decision-making regarding purposeless treatment, thereby offering transparency and reviewability of the whole process.

In the decision-making process regarding the withholding or withdrawal of life-sustaining treatments on the grounds of their purposelessness, attending doctors, the legal guardians (parents) of the patient, and the patient (subject to his/her age and maturity as defined by law) are involved. The wishes of the patient on a treatment which could be taken into consideration in future, must be taken into account. Other attending healthcare workers and the patient's next of kin who show an interest in his/her welfare, are also entitled to express their opinion regarding the treatment. The conclusion of the medical practitioners should be based exclusively on their medical opinion. Although the Paediatrics Guidelines do not explicitly stipulate who would chair the discussion, it is very likely to be the head of department (senior consultant) as he/she is, in principle, involved in all significant and difficult decision-making.

The Paediatric Guidelines also firstly emphasise the empathetic and honest communication between healthcare workers, patients and parents. If the communication is initiated earlier than in the terminal stage and is led systematically, the best possible proposal for the adequate treatment at the end of a patient's life can be implemented. Thus, as a general rule,[37] therapeutic privilege should not be applied. Another positive consequence of honest communication is the prevention of conflicts arising from breaching mutual trust. In this context, systemic establishment and the availability of specialised medical and ethical consultations for the patient and parent/s are also mentioned. Last, but not least, in cooperation with local providers of palliative care and the health insurance company, palliative care which will complement (rather than substitute) curative care should be arranged, including in-patient and out-patient palliative care and grief counselling.

There are four possible situations responded to by the Paediatric Guidelines:

1. Consensus among the persons involved on whether a treatment is purposeless has not been reached. According to the doctors, there is a real benefit for the patient to undergo the treatment. If providing the treatment fulfils the criteria of providing urgent or acute care, doctors shall provide the treatment even without any informed consent.
2. Consensus among the persons involved on whether a treatment is purposeless has not been reached. According to the doctors, the patient should be considered as endangered under section 6 of the Act No. 359/1999 Collection, on the Sociolegal Protection of Children. The sociolegal authorities shall be notified in order to protect the interests of the patient. The patient can be endangered by the fact that the parents are either refusing a necessary treatment or inappropriately demanding intensive and invasive care.
3. Consensus among the persons involved on whether a treatment is purposeless has not been reached. The treatment con-

[37] Despite its legal recognition via section 32 paragraph 2 of the Act on Healthcare Services.

cerned is neither urgent, nor acute (i.e. its provision can be postponed). The patient is also not an endangered child under the law. The court shall be notified in order to decide on the treatment.

4. The patient concerned is fully incompetent, and a highly risky treatment or a treatment with severe side effects is considered. The court shall be notified in order to approve the treatment, even if consensus has been reached among the persons involved on whether the treatment is purposeless.

III. The cases of Charlie Gard and Alfie Evans from the Czech medico-legal perspective

Unlike the Common Law system, Czech Law is codified with more or less merely factual authority of the courts' decisions. In general, a court decision is binding only for the parties/participants in the proceedings or subject to criminal proceedings. However, as the Czech judicial system is three-tier, there is undoubtedly a harmonising role in the judicial practice of the Appeal Courts and especially of the Supreme Court of the Czech Republic.

Until recently, the involvement of a court in a decision-making process regarding medical treatment was mainly limited to the question of the hospital admission of an incompetent patient and his/her treatment in case the legal guardian did not take proper care of the patient.[38] Furthermore, the court was involved when approving a treatment to substitute the refusal of one parent,[39] or when approving a treatment provided to an endangered minor via a preliminary court order.[40] By enact-

[38] The decision of the Czech Constitutional Court, No. III.ÚS 459/03. Available online (in Czech only) at:
http://nalus.usoud.cz/Search/GetText.aspx?sz=3-459-03
(accessed November 29, 2018).

[39] The decision of the Czech Constitutional Court, No. II. ÚS 725/18. Available online (in Czech only) at:
https://nalus.usoud.cz/Search/ResultDetail.aspx?id=104284&pos=1&cnt=2&typ=result (accessed March 13, 2019).

[40] The decision of the Czech Constitutional Court, No. III.ÚS 459/03. Available online (in Czech only) at:
http://nalus.usoud.cz/Search/GetText.aspx?sz=3-459-03
(accessed November 29, 2018).

ing the Civil Code, this scope was significantly enlarged,[41] with no acknowledgment from clinical practice to date. This could be partially explained by the misleading interpretation of the law, as some experts reject the application of the Civil Code to the area of providing healthcare and argue that there is a complex legal regulation in the Act on Healthcare Services.[42] A more practical explanation of the reluctance of healthcare providers to notify the courts might be derived from the traditionally apprehensive position towards authorities, as healthcare providers are unwilling to receive any unnecessary attention.

The Czech Republic has not experienced cases similar to those of Charlie Gard and Alfie Evans. Czech Medical Law does not provide a legal regulation which could offer a definite legal solution, mainly because of the fact that the court's role is not explicitly defined. First of all, the question arises on whether the withdrawal of the mechanical life support of an incompetent patient should be considered as medical intervention (i.e. an active conduct) which is subjected to notification duty under section 101 of the Civil Code, even if doctors and parent/s reach a consensus on it. If concluded that the decision not to provide further medical intervention is in fact *an omission*, and as such it does not require court approval, there is also no explicit legal mechanism for seeking a court order in situations of a dispute between the parents and doctors regarding the withdrawal of life-sustaining treatment.

It is thus extremely alarming that court involvement in such cases can only be assumed, most likely in accordance with section 10 paragraph 1 on the basis of the analogy with section 101 of the Civil Code. In addition, it is also arguable whether the court involved should be a specialised Children's Court instead of a standard court which is to proceed under section 65 and following the Act No. 292/2013 Collec-

---

41 As discussed above, among others, court approval is needed in order to provide healthcare to a minor older than 12 years in case there is a dispute on the treatment between the minor and his/her parents (section 100 of the Civil Code), or in order to provide a highly risky treatment to an incompetent patient (section 101 of the Civil Code).

42 E.g. Petr Lavicky et al, Občanský zákoník I. Obecná část (§ 1 – §654). Komentář, [Civil Code I. General Part (S. 1–S. 654). Commentary], Prague: C. H. Beck, 2014, 579 (available in Czech only).

tion, on Specific Court Proceedings.[43] Further on, it is undecided whether the court proceeding should have the nature of a contentious or non-contentious proceeding.

Last but not least, the criteria by which the court might decide the case are not laid down by the law. Again, only based on analogy with section 102 of the Civil Code, it can be assumed that a court may only approve another life-sustaining treatment if it is, according to their reasonable consideration, beneficial to the patient. Thereafter, the decision should be taken with the full recognition of the personal rights of the patient, after the court has seen the patient.[44] The question whether another life-sustaining treatment constitutes a benefit for the patient is likely to be decided (among others) on the grounds of expert opinion. Despite not being a source of law, guidelines defining medical futility might serve as a useful aid both for the expert and the court whilst confronting their point of view with the current prevailing medical opinion.

Unfortunately, there is no time limit set by the law for the court decision to be made. Within the Czech judicial practice, which often suffers from excessive length of proceedings,[45] this aspect causes further doubts about the whole process of notification within clinical practice.

IV. Conclusion

Czech Medical Law has traditionally held a strongly protective position towards minors' life, health, and access to healthcare. As a consequence, a robust legal mechanism for the treatment of a minor even

[43] Available (in Czech only) at: https://www.zakonyprolidi.cz/cs/2013-292 (accessed November 29, 2018).

[44] As analysed by Tomáš Holčapek, Petr Šustek and Martin Šolc in: T. Holčapek, P. Šustek, M. Šolc, "Odpojení Dítěte od Přístrojů: Na Okraj Případu Charlie Gard" [Disconnecting A Child From Machines. Regarding the Case of Charlie Gard] *Právní rozhledy* 3 (2018): 94 (available in Czech only).

[45] As regards e.g. decisions on guardianship, the median time for court in Chomutov was 435 days in 2017. Data available from the official statistics (České soudnictví 2017. Výroční statistická zpráva) published by the Czech Ministry of Justice. Available online (in Czech only) at: https://www.justice.cz/documents/12681/719244/2017_vyrocni_stat_zprava.pdf/27ba4524-49cb-4744-b834-2c6812f13e5d, p. 59 (accessed March 14, 2019).

without his/her consent or the consent of his/her legal guardian/s has been established which is carefully followed in medical practice. However, with the rapid advances in palliative care in the Czech Republic, discussion has also started regarding the quality of a life sustained by providing often excessively intensive and invasive treatment, resulting in (non-binding) professional guidelines regarding futile and/or purposeless treatment.

Nevertheless, as medical practitioners suffer from legal uncertainty arising from ambiguous legal regulations, they often perform defensive medicine in which the withholding or withdrawal of a life-sustaining treatment, from the legal point of view, is considered overly risky. Unlike Commonwealth countries, it is not standard practice for Czech doctors/healthcare providers to seek a court order. With the exception of seeking approval for the hospitalisation of an incompetent patient, which is done routinely, this is avoided as far as possible. There is no legal experience with cases such as those of Charlie Gard or Alfie Evans, and it is difficult to contemplate whether doctors/healthcare providers would call for the withdrawal of life support in the first place.

In any case, with respect to the whole Czech medicolegal context, court involvement seems to be crucial in situations where life-sustaining treatment is provided to a child and its purposefulness is challenged. As soon as clinical practices start to notify the difficult cases to the court in order to seek a court order, the unsustainable legal unpredictability regarding the end-of-life decision-making based on medical futility might be diminished.

Acknowledgments

This work would not have been possible without the support of the Research Centre for Human Rights (project UNCE 204006 - Research Centre for Human Rights, Charles University).

Chapter 12

The Application of End-of-Life Legislation to Minors in France

Stephanie Rohlfing-Dijoux
University Paris Nanterre, France

I. Introduction

On 2nd February 2016, France passed a new end-of-life legislation[1] (called the Claeys-Leonetti Act) that integrated, for the first time, a ban on unreasonable therapeutic obstinacy, as well as the possibility of terminal sedation. The fundamental principle of this legislation is to allow ill people have their life ended with dignity and be accompanied in their last moments.[2] One of the most important parts of the reform is the strengthening of patients' rights and their constitutionally-granted right to autonomy.

One of the ethical foundations and fundamental motivations for the creation of the Claeys-Leonetti Act was that all patients should receive the best care possible and have their dignity respected[3] at the time their lives end, in accordance with the principles of equity and distributive justice. In France, like in England and Wales, healthcare is accessible and cost-free for every French citizen.[4] Furthermore, French citizens have the freedom to choose their medical practitioners, as well as the

---

[1] Law no. 2016-87 of 2nd February 2016 created new rights for patients and people in their final stages of life.

[2] B. Devalois, L. Puybasset, *Nouvelle Loi Sur La Fin de Vie: Quel Impact pour la Pratique Médicale?* [New Law on End-of-life Situations: What Impact will it have on Medical Practice?], *La Presse Médicale* 4 (2016): 414.

[3] The concept of human dignity in French law is contained in Art. 16-1 and 16-1-1 of the Civil Code and Art. 2 of ECHR.

[4] J. J. Dupeyroux, *Droit de la Sécurité Sociale* [Social Security Law] (Paris: Dalloz, 2015), 468, no. 611; A. Lami, V. Vioujas, *Droit Hospitalier* [Hospital Law] (Paris: Bruylant, 2018), 368.

hospitals for their treatment.[5] Hospitalisation is free of charge in public hospitals, whereas in private clinics, patients would have to prepay the medical fees and then wait for refunds from the public social security service. The refundable costs are decided by social security.[6] The physicians and other medical professionals may ask for overpayment. In this case, the entire costs are not refunded, only the rate decided by the social security, as the rest is charged to the patient. In the case of long and serious illnesses, hospital costs are fully refunded.

The patient is at the core of the new legal framework. All medical decisions have to be taken in his interest and for his welfare, particularly as the Act gives the patient the right to ask for fundamental sedation if no other means can appease pain. The patient may also refuse any medical treatment even if it is necessary to maintain life, or if the means has as its sole aim the artificial prolongation of life without any curative effect. If the patient is not able to express his will anymore, he can do so by advance directives or by designating a person of trust, who will make decisions on the patient's behalf. The patient also has the right to receive all information concerning his state of health or illness in order to be able to make informed and reasonable decisions.[7]

The question is, to what extent can the principles of the new Act be applied to minor patients? Further, can children perform their rights by themselves or do they need the support of parents or legal representatives?

Conflicts may arise if a minor patient and his parents disagree on the biological testament of the minor or if the parents do not agree with the decision of the medical team and thus neither party can proceed with what they perceive as the best medical decision for the child.

This chapter examines these questions by referring comparatively to the recent jurisprudence in France and England and Wales, where similar problems occur.

[5] A. Lami, V. Vioujas, *Droit Hospitalier*, 372, no. 474.

[6] J. J. Dupeyroux, *Droit de la Sécurité Sociale*, 487, no. 642.

[7] S. Rohlfing-Dijoux, "Sterbehilfe in Frankreich: Das neue Gesetz vom Februar 2016" [Assisted Dying in France: The New Law of February 2016], *MedR* (2016): 606.

II. The special situation of minors

Children with serious and incurable illnesses, who are in the final stages of life, are in a unique situation. On the one hand, they need a special reinforced protection, while on the other hand, they do not have autonomy on the fundamental questions concerning their life and dignity. The new French Act contains neither special provisions for minors nor an interdiction of its application to minors. So, it has to be applied in the same way as for adults, but should be combined with the stipulations on minors in family law and minor protection law.

It is difficult to legislate special provisions for minors in end-of-life situations. A person is a minor up until the age of 18 both in France and in England and Wales. Throughout this time, the minor matures and develops his capacity of discernment: growing from a baby unable to express himself verbally into a responsible adult. The different stages of child development should be considered in the legislation concerning the patients' rights.[8]

One should link the right with the child's age. It is impossible to give the same right of self-determination to a small child, unable to understand the consequences of his decision, as to an adolescent who is nearly 18 years of age. A small child could not perform his rights concerning medical treatment. This is the role of his parents or legal representatives. As the child grows, his capacity to decide for himself should increasingly be considered by the legislation.

The necessity to take into account the different stages of child development may be explained by two different cases in France, one concerning a baby and the other a young adult.

In the first case, that of Marwa, a boy under the age of 1, was infected with a rare virus known as *rhombencéphalomyélite enterovirus* which resulted in serious neurological lesions, the paralysis of his arms, legs, and face. He thus became dependent on artificial breathing and nutrition. The baby was for the first time admitted into hospital in September 2016 and was plunged in an artificial coma. The medical team wanted to stop the life-sustaining measures (artificial breathing and nutrition) because, in their opinion, the illness was irreversible and there was no hope that Marwa could return to a "normal" life with a

[8] The Report of 11th June 2018 on the Evaluation of the Application of the Law of 2nd February 2016 on the End-of-life Decisions (IGAS) 83, no. 443.

certain degree of communication and autonomy. The medical decision to stop life-sustaining measures was taken after a collective procedure regulated by Art L. 1110-5-1 of the French Public Health Act (*code de la santé publique*). It was clarified by the by-law of August 2016, a decision which concluded that maintaining life-support was qualified as futile in the sense of Art L. 1110-5-1 of the French Public Health Act (*code de la santé publique*). The parents did not agree with the medical decision and wanted to maintain their child alive at any costs. They submitted their case to the Administrative Court of Marseille in summary proceedings. The court suspended temporarily the execution of the medical decision and ordered for medical expert testimony in an interim injunction of 16th November 2016. In consequence of the expert testimony, in a second decision of 8th February 2017, the court suspended the decision to withdraw the treatments and placed the medical team under an obligation to continue life-stretching measures for the child, at least temporarily. The public hospitals of Marseille (*L'Assistance publique – Hôpitaux de Marseille AP-HM*) gave notice to appeal the decision in front of the judge of summary proceedings of the *Conseil d'Etat* (which is equivalent to the Privy Council).

In a decision of 8th March 2017, the *Conseil d'Etat* decided for the first time by the application of the new dispositions of the Claeys-Leonetti Act, confirming the order of the Administrative Court of Marseille (*TA de Marseille*) and highlighting the "particular importance" of the opinion of the parents, which was "all the more important as it was impossible to search for the will of the child in this case."[9] After the ruling, parental view merits greater weight in the decision-making procedure. In fact, the court considered that not enough time had passed since the first hospitalisation of the child in order to determine whether the treatment was futile. Hence, the *Conseil d'Etat* invalidated the medical team's decision.

The decision-making process and the implementation of end-of-life means (such as sedation in the final phase) in case of children, requires special methods.[10] The collective procedure has to be implemented

[9] *Conseil d'Etat, 8th March 2017*, no. 408146.

[10] M. L. Viallard et al, "Modalités Pratiques d'une Sédation en phase Terminale ou en fin de vie en Pédiatrie: Prise de Décision, Mise en œuvre et Surveillance" [Practical Methods of Sedation in the Terminal Phase or End-of-life in Paedi-

without failure, and the decision has to be anticipated as parents have to be prepared and need time for reflexion. The medical team has to take time to discuss with the child (if he or she is sufficiently mature to engage in the discussion) and the parents, to inform them on the consequences, and prepare them while they take the time to think about it.

The second case concerns an adolescent, 14 year-old Inès, who suffered from a neuromuscular autoimmune illness and was admitted to hospital in June 2017 as a consequence of a heart attack. After different medical treatments, the medical team considered her state of health hopeless and launched a collective decision-making procedure with the aim to withdraw life- sustaining measures according to Art L. 1110-5-1 of the French Public Health Act (*code de la santé publique*).

The situation was nearly the same as in Marwa's case, as the parents wanted to maintain life-stretching measures for their child against the medical opinion of the physicians who considered them to be futile and thus arguably forbidden by the Claeys-Leonetti Act. The parents brought their plea to the court. The *Conseil d'Etat* confirmed the medical team's decision to withdraw the child's artificial support and in a decision of 5th January 2018,[11] it confirmed the decision of the Administrative Court of Nancy of 7th December 2017. The parents appealed to the European Court of Human Rights (ECtHR) in Strasbourg, who supported the decision of the French High Court and the medical team, and declared the decision in accordance with Article 2 of the European Convention on Human Rights.[12] Inès passed away on 21st June 2018 after the withdrawal of medical treatments. According to the Claeys-Leonetti Act, a collective procedure must be conducted before the final decision was made, thus several formal interviews with the parents and relatives had taken place.

These two cases pose two legal questions which will be discussed successively:

- The first question concerns the importance of the child's age when deciding whether a treatment is futile and its withdrawal is needed.

---

atric Cases: Decision-making, Implementation and Monitoring], *Médecine Palliative Soins de Support - Accompagnement - Éthique* 9:2 (2010): 87.

[11] *Conseil d'État*, ordinance of 5th January *2018*, no. 416689.

[12] *Afiri and Biddarri v France* (application no. 1828/18) ECHR 032 (2018).

- The second question concerns the position of the parent as the owner of the right of custody.

A. The importance of age in decision making

It is useful to question the importance of age before the withdrawal of life-sustaining measures.

The cases of newborn babies and neonatal reanimation are different from those of young adolescents due to the children's level of dependence on parents. Three categories of minors can be distinguished: nurslings, children, and adolescents. The difference between them stands in their degree of maturity and consequently in their capacity of discernment and possibility to express their will.[13]

Article 371-1 of the French Civil Code concerning parental authority stipulates that "parents have to associate the decisions on their children's life according to their age and their degree of maturity."[14] Article L.1111-4 of the French Public Health Act (*code de la santé publique*) considers that the consent of the minor has to be sought if the minor is able to express will.[15] According to this disposition, the minor (if their age allows it) can be involved in the medical decisions concerning their end-of-life. Physicians should provide the minor with the necessary information adapted to his capacity of discernment, corresponding to his age.[16] Surprisingly, minors show exceptional maturity in such dramatic situations.

In application of these rules, the French Inspection of Social Affairs (IGAS)'s report on the evaluation of the application of the Act of 2nd February 2016 relating to end-of-life, recommended that adolescents be given the opportunity to draw up advance directives and to assign a person of confidence from the age of 15.[17]

---

[13] The Report of 11th June 2018 on the Evaluation of the Application of the law of 2nd February 2016 on the end-of-life decisions (IGAS) 83, no. 443.

[14] "Les Parents Associent L'enfant aux Décisions qui le Concernent, Selon l'âge et son Degré de Maturité."

[15] A. Lami, V. Vioujas, *Droit Hospitalier*, 324, no. 401.

[16] Ibidem, 339, no. 424.

[17] "The Report of 11th June 2018."

B. The scope of parental rights

In cases of life-sustaining treatment withdrawal from a minor, the parents' rights must always be considered in the decision, as persons in custody of the child. The new French Act does not provide a special procedure, but as the application of the dispositions of the Claeys-Leonetti Act is not excluded, they must also be applied to minors, although minors cannot draft advance directives.[18]

Rather, the general rules of the civil law (*Code civil*) with respect to the right of custody are applicable and devolve an overriding role to the parents in the decision. Hence, these two groups of legal dispositions may conflict with each other. They should be conciliated with the special frame of the end-of-life legislation.

The specific dispositions for minors are stipulated in the by-law of the new Act. Article R. 4127-37-2-1 CSP[19] notes that

> in case of decisions of limitation or withdrawal of treatments concerning a minor or a person of full age under legal guardianship, the physician has to search for the opinion of the party with the right of custody or the legal guardian, except in emergency situations when the consultation of these persons is impossible.

Hence, the physician should always try to obtain the consent of the parents except in emergencies and if the physician cannot agree with the person possessing parental custody.[20]

The parents of the minor give their opinion on the withdrawal of the medical treatment, but this is not a declaration of consent. Although the civil code rules in its Article 371-1 that the consent of the parents is necessary for every medical intervention on a minor, the by-law of the 2nd February 2016 stipulates that the parents only have to give their opinion. It is difficult to find the right balance between the parental rights contained in Art 371-1 of the Civil Code, the interest of

---

[18] B. Adam-Ferreira, *La Qualité de vie en tant que Justification de Décisions de fin de vie en Droit Comparé* [The Quality of Life as Justification for End-of-life Decisions in Comparative Law], Doctoral Dissertation, December 2006, Paris Nanterre.

[19] Issu du Décret n° 2016-1067 du 3 août 2016, JORF n°0181 du 5 août 2016 [Coming from the Decree n° 2016-1067 of August 3rd, 2016, JORF n° 0181 of August 5th, 2016].

[20] A. Lami, V. Vioujas, *Droit Hospitalier*, 372, no. 474.

the minor, and the role of the physician who has to take into account the will of the minor.[21] The *Conseil d'Etat* insists on the fact that the final decision has to be made by the medical practitioner and never by the family.[22] However, the *Conseil d'Etat* also states that the medical practitioner should wait before withdrawing treatments until the parents have had an opportunity to take legal action.[23]

In practice, the opinion of the parents is very important because it seems very difficult for the medical team to decide against the will of the parents. The unconditional respect of the collective decision-making procedure is particularly important in these cases in order to offer the family an important role and consider their extreme distress and emotional load. The particular situation of the age of a child, of the parents, as well as the irreversibility of the decision is considered on a case to case basis by both the jurisdiction of the French *Conseil d'Etat* and the ECtHR.[24]

In 2015, the report of the Commission of Social Affairs in France[25] considered that the opinion of the parties with the right of custody is more than a simple testimony and corresponds to the expression of the person of confidence without according the value of consent. The jurisprudence veered away from the traditional role devolved to the right of custody by the common law. Nevertheless, a clarification of the legal framework would be preferable. The rules could be imbued with solutions adopted in the legislations of neighbouring countries.

III. England and Wales: recent developments

In recent years in England and Wales, several high profile end-of-life cases led to heated debates on ethical and legal questions related to the treatments of minors.

Charlie Gard, an 11 month-old boy suffering from a genetic disease, died in July 2017 after a legal battle between his parents and the hospi-

---

[21] Ibidem, 339, no. 423.

[22] Case of Lambert, *Conseil d'Etat*, 14th February 2014, no. 375081. *Conseil d'Etat* 24th June 2014, no. 375081. *Conseil d'Etat*, 19th July 2017, no. 375081. *Conseil d'Etat*, 31st January 2018, no. 375081. ECHR, No. 46043/14, 5th June 2015.

[23] *Conseil d'Etat*, 5th January 2018, no. 416689 (Dalloz, 2018), 71, note Vialla.

[24] ECHR, judgment no. 46043/14, 5th June 2015.

[25] Report of the *Commission des Affaires Sociales*, 17th Apil 2015.

tal. The case resulted from the hospital's decision to take the child off life support, deeming any treatment futile.

Young cancer patient Ashya King was kidnapped by his parents from the hospital where he was being treated (and which also declared his treatment futile) and taken to Prague for a treatment with protons,[26] which had been refused to him in England. The child is currently 8 years old and was declared cured.

The most intricate thus far seems to be the case of Alfie Evans where a particularly violent legal battle with important media coverage was fought by the parents of the 23-month-old boy until his death, on 28th April 2018. Alfie Evans, born on 9th May 2016, was admitted to the Alder Hey Paediatric Hospital in Liverpool with a rare neurodegenerative pathology for which no effective treatment existed. His medical team concluded that it was not in Alfie's interest to continue treatment. The parents formally opposed the medical decision to withdraw life-sustaining treatment and took legal action. The High Court, the Court of Appeal,[27] and the Supreme Court[28] all decided in favour of the medical team. The parents then submitted the case to the ECtHR, but three ECtHR judges considered the request inadmissible because there was no human rights violation.

A large social media mobilisation took place. The parents obtained the support of foreign countries, notably Italy and Poland, and wished to transfer the child to a specialised hospital in Rome. The hospital in Liverpool did not consent to Alfie's transfer arguing that there was a strong risk that the child could die during the journey. Italy granted Alfie Italian citizenship in order to facilitate his transfer. The parents argued for their right of custody, their right to decide the whereabouts of their son, as well as their right to decide on the continuation of medical treatment indispensable to life (health care for the child). All were refused to them by the English courts.

The medical team considered that due to the grave degradation of Alfie's brain cells, maintaining life-stretching measures would not only be a therapeutic obstinacy and a medically futile measure, but also a cruel and inhumane treatment. Nevertheless, the parents continued to request the prolongation of his life at any cost. The English courts

---

26 *Re King* [2014] EWHC 2964 (Fam).

27 Court of Appeal on 6th March 2018.

28 Supreme Court on 20th March 2018.

declared that the matter had to be decided in the best interests of the child if no conciliation between the parents and the physicians could be found.

The principal legal question raised in this case was whether the physicians or the parents are competent to decide if the withdrawal of the treatment is in the best interests of the child. In England and Wales, the legal stipulations of the 1989 Children's Act state that if a child is hurt, the state has to intervene. This means that the parents' rights are not absolute and the judiciary may intervene if they considered that the parents' decisions are not compatible with the best interests of the child. If a public institution, in this case the public hospital, does not agree with the decision of the parents, it has to submit the case to the court for a decision.

All these cases have in common the conflict between parents and the medical team on the question of what is considered medically effective and medically futile. The contentious nature of "medical futility cases" results from the fact that both sides claim to act in the best interests of the child. Thus both sides' behaviour is sustained by good faith. This suggests that the concept of the best interests of the child is not absolute. For some, the only interest of the child is the continuation of life, while others deem it useless to inflict suffering on a child with no hope of cure and of regaining a dignified and autonomous life. Thus the same medical intervention can be considered both detrimental and beneficial. These different cases reveal the varied ways in which "best interests" can be interpreted. Judgment is subjective and depends on the position of the observer (parents and family, or physicians) as both parties are convinced they were acting in the best interests of the child.

IV. A comparative analysis

The solution of the English Courts respects the role attributed by law to custody. There is no conflict between the civil law rules related to custody and case law in England and Wales. Both are in harmony since they can be applied simultaneously. However, in France, the application of the civil law rules on custody (Article 371-1 of the Civil Code) is not compatible with the rules of the public health code (Article L. 1111-4 of the Public Health Code), which determines that on the one hand, the consent of the parents is necessary for every medical intervention on a minor, but on the other hand, the consent of the parents is not mandatory in case of a definitive and irreversible decision of withdrawal of life-sustaining measures. The parents' testimonial is only necessary for the collective medical procedure but without any binding effect. The physician in charge of the patient can and has to decide last and eventually against the will of the parents and without any jurisdictional authorisation.

This way of settling the conflict between the common civil law rules and the special rules of the by-law related to the Public Health Code in favour to the latter are debatable (contestable). An intervention of the French legislation in order to clarify these important questions is preferable and seems necessary in order to avoid other litigious cases. Even if the end–of-life rules are at a crossroads between Family Law and Public Health Law, as the latter is more significant, it should have priority. In this context, the solution of French law is not entirely satisfactory. One may ask whether the parents' opinion should not have a more significant role in the decision-making process in France. Interestingly, the same question could perhaps be asked about English Law.

One of the possibilities which aim to avoid conflicts between the parents of a minor and the medical team could stand in mediation. The report of the French Inspector for Social Affairs (*Inspection Générale des Affaires Sociales -IGAS*)[29] called for the use of mediation in all cases of medical futility on minors in the future. Mediation could play an important role in general, and in particular in cases of minor patients. Mediation could become systematic and mandatory, but the problem is whether parents should be sanctioned if they refuse to participate. In such situations, it is impossible to foresee any enforcing measures such as the refusal to pay medical fees. Therefore, it remains at the discretion of each party to accept it or not. This proposal has also been made by the English courts in the Charlie Gard case.[30]

[29] "The Report of 11th June 2018", recommendation no. 22.

[30] *Great Ormond Street Hospital v Yates, Gard and Gard* [2017] EWHC 1909 (Fam), para. 20.

Chapter 13

Patient Autonomy and Best Interests in End-of-Life Cases: A German Perspective

Peter Elsner
University Hospital Jena, Germany

I. Introduction

Medical staff should act in the best interests of the patient. In doing that, the autonomy of the patient has to be respected. Medical procedures require informed consent and shared decision-making. In the case of persons unable to give informed consent, conflicts may arise between a patient's legal representatives and medical staff on deciding which procedures would be in the patient's best interests.

The highly-publicised case of Charlie Gard led to medico-ethical and legal discussions on how conflicts concerning the best interests of a patient in futile treatment cases should be solved. Charlie was a male child born in August 2016. In October 2016, a diagnosis of infantile onset encephalomyopathic mitochondrial DNA depletion syndrome was made, a rare and fatal genetic disorder linked to a mutation in the RRM2B gene.[1] While the physicians attending Charlie at the Great Ormond Street Hospital (GOSH) submitted that further treatment was futile, the parents wished to take Charlie to the United States for an experimental treatment that had neither been tested on animals, nor on humans-[2] In the High Court, the best interests test was applied, and Mr Justice Francis granted the declarations sought by GOSH - it was not in Charlie's best interests to have nucleoside therapy, and therefore it was lawful to withdraw artificial ventilation and provide palliative care only.[3]

---

[1] N. Hammond-Browning, "When Doctors and Parents Don't Agree: The Story of Charlie Gard," *Journal of Bioethical Inquiry* 14 (2017): 461.
[2] Ibidem.
[3] *Great Ormond Street Hospital v Yates, Gard & Gard* [2017] EWHC 972 (Fam).

This decision was confirmed by the Court of Appeal,[4] by the Supreme Court,[5] and finally by the European Court of Human Rights (ECHR).[6] As Hammond-Browning stated,[7] these decisions are a clear application and endorsement of well-established legal principles according to English law to determine if medical treatment is in a child's best interests, at each level the courts taking into account medical advice, parental wishes, the child's current and future positions, and the possibility of experimental medical treatment. In that regard, the decisions were legally unremarkable.

From an ethical point of view, however, several aspects of the case require further attention: the conflict between parental autonomy and the judgment of medical staff on the best interests of a patient, ways to resolve such conflicts outside the courts, i.e. mediation, and the ethical and legal status of experimental treatments.[8] In addition, considering the media attention, but also the involvement of the American President and the Pope in the case, it has to be discussed how helpful such an environment of hysteric public debate might be for resolving a very personal conflict.

The case of Charlie Gard and the similar cases of Alfie Evans[9] and Ashya King[10] have been discussed in English jurisprudence[11] regarding the best interests test and the right to experimental treatment, but their essence is of universal relevance for medical ethics and medical law. It might therefore be worthwhile to add a continental, especially German, perspective to the legal questions raised in these cases.

---

[4] *Yates and Gard v Great Ormond Street Hospital & Gard* [2017] EWCA Civ 410.

[5] *In the Matter of Charlie Gard* (19th June 2017).

[6] *Gard and Others v the United Kingdom* (application number 39793/17).

[7] Hammond-Browning, "When Doctors and Parents Don't Agree."

[8] I. Brassington, "Charlie Gard: An Ethical Analysis of a Legal Non-Problem," *Journal of Medical Ethics Blog* (11th August 2017). https://blogs.bmj.com/medical-ethics/2017/08/11/charlie-gard-an-ethical-analysis-of-a-legal-non-problem/ (accessed November, 14 2018).

[9] *Alder Hey Children's NHS Foundation Trust v Evans, James & Evans* [2018] EWHC 308 (Fam).

[10] *Re King* [2014] EWHC 2964 (Fam).

[11] "'Best Interest' Cases," *Counsel* (May 2018), https://www.counselmagazine.co.uk/articles/%E2%80%99best-interest%E2%80%99-cases (accessed November 15, 2018).

II. Medical futility and patient autonomy

Medical futility has been defined as the unacceptable likelihood of achieving an effect that the patient has the capacity to appreciate as a benefit.[12] This implies a quantitative aspect, i.e. a statistical probability that needs to be defined, and a qualitative aspect, i.e. the fulfilment of the patient's wishes for the success of a treatment. If the best outcome physicians can achieve to maintain survival requires keeping the patient perpetually confined to the Intensive Care Unit or the acute care hospital setting, this would be considered a qualitatively futile treatment.[13] This concept of futility has been criticised early on[14] with the arguments that (1) the cutoff point of the quantitative, statistical aspect may be arbitrary, (2) the probability of treatment success in a specific case may be difficult to judge, and (3) the qualitative aspect of the definition is a question of values and treatment goals. If a patient wishes a specific kind of treatment, arguably this treatment cannot be considered futile, although society may disregard the wish of the patient.[15] On the other hand, suggestions to use the terms "potentially inappropriate" or "inappropriate" treatment instead of "futile treatment," have been rejected on the grounds that these terms would be vague and paternalistic.[16]

In Germany, the term "futility" has not been used and discussed as in the US and England and Wales. Instead, the focus has been on the medical indication of a given treatment and on the wish of the patient after informed consent (patient autonomy). This was enacted by the 3rd Act on Amendment of the Care Law that came into effect on September 1, 2009[17] which focuses on patient directives (living wills). An adult with the capacity to give informed consent may give or

[12] L. J. Schneiderman, "Defining Medical Futility and Improving Medical Care," *Journal of Bioethical Inquiry* 8 (2011): 123.

[13] L. J. Schneiderman, N. S. Jecker, A. R. Jonsen, "Medical Futility: Its Meaning and Ethical Implications," *Annals of Internal Medicine* 112 (1990): 949.

[14] R. D. Truog, A. S. Brett, J. Frader, "The Problem with Futility," *The New England Journal of Medicine* 236 (1992): 1560.

[15] Ibidem.

[16] L. J. Schneiderman, N. S. Jecker, A. R. Jonsen, "The Abuse of Futility," *Perspectives in Biology and Medicine* 60 (2018): 295.

[17] "Gesetz zur Änderung des Betreuungsrechts," Bundesgesetzblatt Jahrgang 2009 Teil I Nr. 48, ausgegeben zu Bonn am 31. July 2009 2287.

withhold his consent in writing for the event of his incapacity to consent to examinations of his state of health, medical treatment, or medical intervention that are not yet imminent at the time of writing the patient directive (§ 1901a para. 1 of the Civil Code (*Bürgerliches Gesetzbuch,* BGB)). The legal guardian of a patient has to check whether these determinations apply to the current life and treatment situation of the patient. If this is the case, the caregiver has to express the will of the person being cared for and to enforce the law. A patient directive may be revoked at any time, also informally. If no patient directive is available or if the provisions of a patient directive are not met in the current situation of life and treatment, the legal guardian has to identify the treatment wishes or the presumed will of the person being cared for, and, on this basis, he has to decide whether he consents to a medical measure or not (§ 1901a para. 2 BGB). The presumed will has to be determined on the basis of concrete evidence, in particular, earlier oral or written statements, ethical or religious beliefs, and other personal values of the person being cared for. These regulations apply independently of the type and stage of a disease of the person to be cared for (§ 1901a para. 3 BGB). § 1901b BGB refers to a conversation to determine the will of the patient. The attending physician shall examine which medical measure is indicated in view of the overall condition and the prognosis of the patient. He and the legal guardian discuss this measure taking into account the patient's will as a basis for the decision to be made according to § 1901a. In determining the will of the patient, his treatment wishes or his presumed will, the opportunity to comment shall be granted to close relatives and other persons of trust, provided that this is possible without considerable delay (§ 1901b para. 2 BGB). § 1904 BGB states that the approval of the court is required for medical measures. The guardian's consent to an examination of the patient's state of health, medical treatment, or medical intervention requires the approval of the court if there is a well-founded danger implying that the person being cared for dies as a result of the measure or suffers a severe and prolonged damage to his health. Without the consent, the procedure may only be performed if a postponement would cause danger to the patient's life or lead to severe long-lasting damage.

If there is no effective advance directive according to § 1901a para. 1 BGB, and there is no agreement between the legal guardians and the attending physicians that the termination of life-sustaining measures corresponds to the actual or presumed will of the patient established in accordance with § 1901a BGB, the permission of the court is required for the decision.[18] The evidential requirements for the determination of the patient's actual or presumed will are high due to the high value of the legally protected rights.

In a landmark decision handed down in 2010, the German Federal Court (BGH), the supreme court for criminal cases, allowed the appeal of a solicitor who had been held culpable of manslaughter since he had advised his client to cut the tube used for artificial nutrition, an act regarded as active assisted dying by the lower court.[19] The BGH stated that assisted dying by omitting, limiting, or terminating a medical treatment is justified if it is in accordance with the patient's actual or presumed will (§ 1901a BGB) and if it serves to allow a disease process leading to death without treatment to run its course, and that such a termination may be performed by omission as well as action.[20]

III. Parental care and medical futility

Medical futility cases relating to children are legally different from those on adults. While the actual or presumed wish of the patient is paramount in guiding a decision on medical interventions in adults, children are not legally capable of deciding; thus they need to be represented by adults, i.e. primarily by the parents. The rights and duties of the parents to take care of their children are enshrined in the German Constitution (Basic Law, Grundgesetz) in Art. 6 para. 2: "The care and education of children are the natural rights of parents and their primary duty. Their activities are supervised by the state." Parental rights and duties are detailed in the Civil Code (BGB) § 1626 para. 1: "Parental custody shall include custody of the person of the child (personal care) and the property of the child (property care)." Thus, parents have a constitutionally guaranteed fundamental right to make a decision in favour of their child according to their own parental "world

[18] BGH, Decision of 17 September 2014 – XII ZB 202/13 –.

[19] BGH, Judgment of 25 June 2010 – 2 StR 454/09 –.

[20] Ibidem.

of values."[21] They may decide on the treatment of their child in the same way as they would decide on their own treatment. This is the core content of the basic right of Art. 6 Grundgesetz (GG), protecting the values of the family as much as possible.[22]

However, parental custody is not unlimited. If the physical, mental, or emotional well-being of the child or his or her property are endangered and the parents are unwilling or unable to avert the danger, the family court has to take the measures which are necessary to avert the risk (BGB § 1666 para. 1). The measures have to be proportionate. The court may cancel or modify the declarations of the holder of parental authority, or partially or totally withdraw parental authority.

For each medical intervention on a child, the consent of both parents is necessary. If it is foreseeable that the child will suffer harm where parental consent was refused for a medically necessary procedure, the treating physician must assert himself against the parents in the child's interests. He may involve the family court to assist in such an action. For example, if the parents refuse to continue the chemotherapy of a child, which would make a difference between life and death, the treating physicians do not act illegally if they bring this to the attention of the family court and suggest measures according to §1666 BGB.[23] If the child's life is acutely threatened, the family court may issue an urgent decision. In even more urgent cases, the physician may treat this as a supra-legal state of emergency and proceed as deemed necessary, but only if such medical action is the only way to avoid irreversible harm to the child.

Thus, the fundamental right of the parents to decide on the treatment of their child finds its limits only in the objective welfare of the child (best interests). The decision on terminating life-prolonging measures is also included in the scope of parental responsibility,[24] and a parental decision, if it is in the child's best interests, has to be respected by the court. The Appellate Court (OLG) of Hamm had to decide on

---

[21] OLG Hamm 1 Senat für Familiensachen, Decision of 24.05.2007 - 1 UF 78/07.

[22] Ibidem.

[23] Oberlandesgericht des Landes Sachsen-Anhalt, Judgement of 12.09.2013 – 1 U 7/12 –.

[24] Brandenburgisches Oberlandesgericht, Judgement of 17.02.2000 – 10 UF 45/99 –.

an appeal to the decision of the district court that had withdrawn the parental authority since the parents had decided not to agree to continue life-prolonging treatment in the case of a 4-year-old child with irreversible apallic syndrome without brain function. The court stated:

> Now all jurisprudence on abortion of treatment, admittance of dying and observance of a patient directive shows that it has long since ceased to apply that admitting dying *per se* is contrary to the well-being of a human being.[25] ... The proceedings are not about the state deciding whether the life of such a suffering coma patient is worth living or not worth living. The only issue in these proceedings is whether the parents' decisions to let the child die have exceeded their constitutionally protected decision-making rights.[26]

IV. Experimental treatments

All innovative medical therapies have started as "experimental treatment"; i.e. there may be *in vitro* or animal experiments suggesting the treatment to be effective and safe in humans, but the experiments are yet to be proven by clinical studies, and the treatments have not yet been approved by regulating authorities, e.g. drug administrations. The legal questions which condition such experimental treatments on humans are manifold. They refer to the legal requirements for clinical trials, but also to the right of the insured patients to receive (often very costly) compensation for experimental treatments from their insurance companies or state-run health services like the NHS in the UK when the procedure goes wrong and causes harm, as well as the right of patients and legal guardians to opt for experimental therapies for children or patients legally incapable of decision-making.

In Germany, most patients are insured by the social health insurance, with the social insurance law in the Social Code V (*Sozialgesetzbuch V*) to be applied. According to §12 SGB V, the services to be supplied must be "sufficient, appropriate and economical; they must not exceed what is necessary."

---

25 OLG Hamm 1. Senat für Familiensachen, Decision of 24.05.2007 - 1 UF 78/07.

26 Ibidem.

An experimental therapy (e.g. an off label use of an approved drug) can only be considered in principle to be covered by the German social health insurance if it concerns (1) the treatment of a serious (life-threatening or impairing the quality of life in the long-term) condition, if (2) no other therapy is available, and if, on the basis of the data available, (3) there is a reasonable expectation that with the treatment in question a curative or palliative treatment success can be achieved.[27] This decision of the Federal Social Court (*Bundessozialgericht*, BSG) was an adoption of a decision of the Federal Constitutional Court (*Bundesverfassungsgericht*, BVerfG) on December 6, 2005 (the so-called "Santa Clause judgement"). It stated that it is not in accordance with the fundamental rights of Art. 2 (1) of the Basic Law to exclude a person covered by the statutory health insurance (for whose life-threatening or regularly fatal disease a generally recognised medical treatment is not available) from a method of treatment chosen by him and applied by a physician, if there is a not entirely distant prospect of healing or of a noticeable positive effect on the course of the disease.[28] The German Parliament reacted to these judicial developments with the addition of § 2 (1a) SGB V that came into force on January 1, 2012 and that codifies the constitutional requirements for the compensation of experimental therapies by the social health insurance.

Thus, in a case comparable to that of Charlie Gard (which has never been brought before the German courts so far), the parents' demand for experimental treatment to be funded by the German social health insurance would have to be judged according to the above three criteria developed by the courts and now codified in social health law. The first two conditions, i.e. a life-threatening disease and the non-existence of a generally recognised medical treatment, are clearly satisfied here. However, it is a question of judicial discretion if the third condition (i.e. the experimental therapy should provide a "not entirely distant prospect of healing or of a noticeable positive effect on the course of the disease") is also met. Answering this question of medical science and expertise would most probably require the opinion of independent medical experts on the order of the courts.

[27] BSG, Judgement of 04.04.2006 – B 1 KR 7/05 R.

[28] BVerfG, Decision of 06.12.2005 – 1 BvR 347/98.

Considering the high value of patient autonomy in German law which is based on constitutional rights as outlined above, a patient is free to choose an experimental therapy not covered by the social health insurance, if the above criteria are not fulfilled, under the condition that he covers the costs of treatment himself. The same would probably apply to the decision of parents to consent to such a therapy for their child, if they covered the costs themselves and if the treatment would not impair the objective welfare of the child (best interests). Circumstances which would compromise the child's best interests include where an effective and generally recognised treatment available was rejected in favour of a doubtful experimental therapy or that the experimental treatment would lead to additional and probably unnecessary suffering of the child. In any case, such a parental treatment decision which might possibly be in conflict with the advice of treating physicians would have to be judged by the family court on the basis of BGB § 1666 para. 1.

Chapter 14

Gard and Evans: A Reflection on What Might Happen in India

Abhay Vaidya,[1] Sourabhi Sahakari[2]
[1] University of Central Lancashire, UK
[2] G R Kare College of Law, India

I. Introduction

In this chapter, we enumerate the challenges Indian parents could face if they had a child with a terminal untreatable medical condition such as that of Charlie Gard ("Charlie") and Alfie Evans ("Alfie"). We comment on what role medical ethics (or lack of it) plays in treating patients, as well as on the challenges, including legal, faced by the medical profession.

The first part of this chapter will describe the healthcare model of India that is probably not significantly different from the healthcare model in many other developing countries across the globe but which differs significantly from that in the United Kingdom (UK). This should facilitate the reader to understand the challenges faced by an ordinary Indian parent.

In the second part, we expand on the law as applicable and interpreted by the courts in India which could have influenced the care given to similarly-situated children in India. Although Indian Law has its roots in English Law, we argue that in many respects (viz. suicide, passive euthanasia) Indian Law has failed to keep pace with the current opinions in the world at large, therefore cases like Charlie's and Alfie's would have been treated differently in India. We believe that the two cases are the distinctive product of their environment.

II. The Healthcare Systems

A. The United Kingdom

In the UK, the NHS became operational on 5th July 1948 and was the brainchild of the Minister of Health Aneurin Bevan.[1] The NHS provides free healthcare at the point of need to all UK residents regardless of wealth[2] and is funded through general taxation. This model of the NHS has survived, largely unchanged, for the last 70 years and the NHS has been highly successful in caring for the UK population. For 2015-2016, the overall NHS budget was around £116.4 billion[3] whereas the total commissioning budget for 2019-20 is £121 billion.[4] The NHS employs nearly 1.5 million people of which 1.2 million are employed in England where a population of 54.3 million is served by 150,273 doctors.[5] In the NHS, patients do not choose the hospital or the place of the treatment, but are rather managed in a facility most suitable to meet their clinical needs.

Charlie and Alfie received treatment in two of the best known and world-renowned NHS paediatric hospitals in the UK – London's Great Ormond Street Hospital (GOSH) and Liverpool's Alder Hey Children's Hospital respectively. Despite the austerity the NHS is facing, it has not been suggested that medical decisions were either compromised or influenced by financial or other resource constraints. The doctors treating both infants followed the General Medical Council's (GMC) ethical guidance Good Medical Practice (GMP) 2013[6] which states the following: "[m]ake the care of your patient your first concern" and "[w]ork in partnership with patients." Regrettably, it appears that in

---

1 "Aneurin Bevan," http://www.nyebevan.org.uk (accessed July 18, 2018).

2 "About the NHS," https://www.nhs.uk/using-the-nhs/about-the-nhs/the-nhs/ (accessed August 19, 2018).

3 Ibidem.

4 NHS England and NHS Improvement Funding and Resource 2019/20: Supporting 'The NHS Long Term Plan' (2019), https://www.england.nhs.uk/wp-content/uploads/2019/03/nhse-mhsi-funding-and-resource-2019-20-supporting-nhs-ltp.pdf (accessed June 1, 2019).

5 "About the NHS."

6 General Medical Council, "Good Medical Practice," (2013). https://www.gmc-uk.org/ethical-guidance/ethical-guidance-for-doctors/good-medical-practice (accessed July 18, 2018).

both cases, communication and partnership between doctors and parents broke down at a certain stage. Both sets of parents disagreed with the treating physicians regarding what they perceived as the best interests of their sons. They argued for continuation of the medical care and challenged doctors' decision of not allowing them to take the infants overseas for further treatment.[7] Both infants continued to be treated in the intensive care units until the courts decided that palliative care was in their best interests.

B. India

The Indian healthcare system will now be discussed in more detail so as to highlight how it differs not only from the fully state-funded NHS, but also from fully and/or partially insurance-funded systems that exist in different parts of the world.

B.1. Healthcare provision

India is a federalised system with 29 states and seven Union territories. The Union Ministry of Health and Wellbeing is responsible for certain national programmes such as training and standard setting for various health educational programmes, food safety and standards, control of diseases such as AIDS or Tuberculosis, to name a few, and so on.[8] The state governments are responsible for public health, hospitals, sanitation, etc.[9] with each state government having its own laws relating to healthcare.[10] According to the National Health Profile report (12th

[7] K. Phillips, L. Bever, "Charlie Gard's Parents to Present New Evidence in Case Surrounding Terminally Ill Son," *The Washington Post* (July 10, 2017). https://www.washingtonpost.com/news/worldviews/wp/2017/07/09/he-deserves-a-chance-charlie-gards-parents-fight-to-bring-son-to-the-u-s-for-treatment/?utm_term=.0ffed4aa7839; "Alfie Evans Parents Appeal Against Italy Travel Ban Ruling," *BBC* (April 25, 2018). https://www.bbc.co.uk/news/uk-england-merseyside-43883865.

[8] M. Chokshi et al, "Health Systems in India," *Journal of Perinatology* 36:Suppl 3 (2016): S9–S12. https://www.ncbi.nlm.nih.gov/pmc/articles/PMC5144115/.

[9] Ibidem.

[10] S. Kar, "Addressing Underlying Causes of Violence Against Doctors in India," *The Lancet* 389:10083 (1979-1980). https://www.thelancet.com/journals/lancet/article/PIIS0140-6736(17)31297-7/fulltext.

edition), for a population of 1.21 billion, there are 14,379 government hospitals with 634,879 beds of which 425,869 beds are in urban areas and 209,010 in rural areas although approximately 70% of the Indian population live in the rural area.[11] The Indian healthcare system is a complex mixture of allopathic[12] and several traditional (complementary)[13] medicines.[14] The Government of India's Ministry of Health and Family Welfare became responsible, amongst other things, for allopathic medicine when the Department of Ayurveda, Yoga and Naturopathy, Unani, Siddha and Homoeopathy (AYUSH) was upgraded to the Ministry of AYUSH in 2014 in order "to ensure the optimal development and propagation of AYUSH systems of health care."[15] In 2016, the total number of allopathic and AYUSH doctors in India was 1,005,281 and 771,468 respectively.[16] Currently, AYUSH practitioners are not licensed to practise allopathic medicine.[17] In 2017, a bridge course was proposed that would have allowed AYUSH practitioners and nurses to prescribe essential (allopathic) medicines to improve healthcare access in rural India.[18] However, this proposal was not accepted by the Government of India in the face of opposition mainly from the allopathic practitioners and the Indian Medical

---

11 "National Health Profile," 12th Edition, *Central Bureau of Health Intelligence.* http://www.indiaenvironmentportal.org.in/files/file/NHP_2017-1.pdf (accessed August 25, 2018).

12 Allopathic medicine is also known as western or modern medicine. See Legal Status, WHO, 2001.

13 The Indian traditional medicine, known as AYUSH, is comprised of Ayurveda, Yoga and Naturopathy, Unani, Siddha, Homeopathy and Sowa-Rigpa.

14 World Health Organisation, "Legal Status of Traditional Medicine and Complementary/Alternative Medicine: A Worldwide Review (2001). D. Shankar, B. Patwardhan, "AYUSH for New India: Vision and Strategy," *Journal of Ayurveda and Integrative Medicine* 8:3 (2017): 137–139.

15 Ministry of AYUSH. http://ayush.gov.in (accessed August 25, 2018).

16 "National Health Profile."

17 A. Marthanda Pillai, K. K. Aggarwal, "AYUSH Cannot Prescribe Modern Medicine Drugs," *Indian Medical Association,* http://www.ima-india.org/ima/left-side-bar.php?scid=245 (accessed September 5, 2018).

18 A. Jyoti, "Getting AYUSH into Modern Medicine," *The Pioneer* (July 25, 2017), https://www.dailypioneer.com/columnists/oped/getting-ayush-into-modern-medicine.html.

Association.[19] The precise distribution of beds in allopathic and AYUSH hospitals is unclear from the above cited report. Most of the rural population and financially disadvantaged urban people rely entirely on the government/charitable institutions for access to healthcare whereas those better off are able to entirely fund themselves.[20] [21] [22]

The government funding for healthcare has significantly declined since Independence in 1947[23] and current healthcare spending is reported to be just above 1.0% of the GDP, making it one of the lowest in the world.[24] Of the total health spending, the government only contributes 29% in India compared to 83% in the UK.[25] This has resulted in India having one of the world's highest out-of-pocket spending on healthcare.[26] Due to a lack of widespread health insurance, personal funding constitutes nearly 62.6% of healthcare spending, far more than most other BRICS countries, the UK, and the USA.[27] In 2013-2014, government insurance covered only 12.78% of Indians whereas private insurance covered only 2.47% of the population.[28] In September 2018, the Government of India launched a universal healthcare system which entitled about 500 million people living on low income to free health insurance and treatment in both government and

---

19 D. N. Jha, "Cabinet Junks Bridge Course Proposal for AYUSH Doctors," *The Times of India* (March 29, 2018).

20 R. Almeida, J. Benrey, J. Freimark, M. Houck, L. Liu, M. Messing, A. Tan, A. Weiss, "The Role of Government in the Indian Hospital System," Woodrow Wilson School of Public and International Affairs, Princeton University. http://wws.princeton.edu/sites/default/files/content/docs/India%20Workshop%20Report_2017.05.10%20FINAL.pdf (accessed 25 Aug. 2018).

21 A. Sengupta, S. Nundy, "The Private Health Sector in India: Is Burgeoning, but at the Cost of Public Health Care," *British Medical Journal* 331:7526 (2005): 1157-1158.

22 S. Dey, "India's Health Spend Just Over 1% of GDP," *The Times of India* (June 20, 2018).

23 R. Almeida et al, "The Role of Government in the Indian Hospital System."

24 S. Dey, "India's Health Spend Just Over 1% of GDP."

25 S. Bansal, "How India Spends on Health," *Hindustan Times* (December 11, 2017).

26 Ibidem.

27 Ibidem.

28 R. Almeida et al, "The Role of Government in the Indian Hospital System."

private hospitals. However, there are concerns that the scheme has been rushed and is poorly funded.[29] Its effectiveness is yet to be judged.

Paying privately or out-of-pocket for medical treatment in a facility of personal choice is a common practice in many developed as well as developing countries of the world. Patients in India often pay from their own pockets for treatment[30] because healthcare is fragmented[31] and many government-run healthcare facilities (including hospitals) are ill-equipped and poorly resourced.[32]

In this chapter, we will mainly compare allopathic healthcare systems for ease of comparison with the NHS. The structure of healthcare delivery discussed below is probably not unique to India but applicable to many other developing countries worldwide.

B.2. Health service facilities

The Indian healthcare facilities can be broadly divided into several layers which often overlap.

B.2.1. Private practitioners

These are qualified "self-employed" medical practitioners. They often act as general/family practitioners (GPs). Unlike the NHS primary care structure where every individual must register with a GP, there is no such obligation in India and patients choose their own doctor when they need one. Patients pay in full for their consultation and treatment with the doctor. A patient needing secondary care would be referred to a hospital of the patient's choice which may depend on the patient's ability to pay as well as local access/availability. Alternatively, patients

---

29 A. Withnall, "Modicare: India Launches World's Biggest Experiment in Universal Healthcare," *Independent* (September 22, 2018).

30 R. Almeida et al, "The Role of Government in the Indian Hospital System"; A. Sengupta, S. Nundy, "The Private Health Sector in India."

31 World Health Organisation, "Legal Status of Traditional Medicine." D. Shankar, B. Patwardhan, "AYUSH for New India."

32 A. Sengupta, S. Nundy, "The Private Health Sector." S. D. D'Cunha, "Despite a Booming Economy, India's Public Health System is Still Failing its Poor," *Forbes* (September 12, 2017), https://www.forbes.com/sites/suparnadutt/2017/09/12/despite-a-booming-economy-indias-public-health-system-is-still-failing-its-poor/#2320e1378e0a.

may access secondary care directly by bypassing a GP.

B.2.2. Small private hospitals

These are small hospitals owned and operated by individual doctors often with postgraduate qualifications. The number of beds in such single-doctor-managed hospitals may vary between 1 to 25. Some larger hospitals with 25 to 100 beds may be single or multiple-doctor-managed. The costs of consultation and treatment are fully paid for by the patient. Most hospitals would not entertain high risk cases as they may lack resources.[33]

B 2.3. Charitable[34]/Trust[35]/Missionary[36] hospitals

These hospitals provide secondary care. Bed capacity may vary from 100 to 250 plus.[37] Treatment provided could be free or paid.

B.2.4. Corporate[38] hospitals

These are multi-specialty hospitals often known as 5* hospitals. Some have a "research centre" status. As research centres, these hospitals benefit from many government subsidies such as exemptions from taxes and duties on import of medical equipment, drugs, and prime land at subsidised rate.[39] One of the conditions of being designated as research centre is that a quarter of the patients are treated free, but this

---

33 A. Kate, "Future of Small Hospitals in India," in *Medicine Update,* ed. A Muruganathan (Association of Physicians of India), Chapter 155, vol. 23 (Delhi: Jaypee Brother Medical Publishers, 2013), 686-687.

34 Bhatia Hospital Mumbai; Nair Hospital Mumbai; Dhanwantari Hospital and Research Centre Mumbai; Karnatak Health Institute, Ghataprabha. Note that the list of hospitals referred to in this chapter is not exhaustive. The first author was trained and has worked in some of the hospitals named.

35 Tata Cancer Hospital Mumbai; Medical Trust Hospital Kochi.

36 Christian Hospital Mungeli; Bangalore Baptist Hospital, Bangalore; Christian Medical College, Ludhiana; Ramkrishna Mission Sevashrama; Ramakrishna Mission Hospital, Mumbai.

37 A. Kate, "Future of Small Hospitals in India."

38 Apollo hospitals in several cities; Jaslok Hospital and Research Centre Mumbai.

39 A. Sengupta, S. Nundy, "The Private Health Sector in India."

condition is rarely met.[40] Most patients pay. These hospitals employ leading medical practitioners and have some of the latest technologies. Services are charged at a (steep) price.

B.2.5. Private[41] medical college hospitals

In India, many private institutions operate medical colleges (medical schools) for full fee-paying students and often have hospitals attached to facilitate training of their students. The treatment may be free, subsidised, or charged at full commercial rate.

B.2.6. Government hospitals

These are run either by local (municipal) authorities[42] or the state governments.[43] The services provided may vary from primary care centres, small cottage hospitals, district hospitals to larger hospitals (some of which may even have a medical college attached). These hospitals are often burdened by government bureaucracy and have a reputation for overworked and overburdened doctors and staff, lack of investigative facilities, dirty surroundings, and many are known to harbour corrupt practices relating to, for example, access to senior doctors, clean bed sheets, and even better food.[44] It is therefore not surprising that although all the care (including medicines) are free in these hospitals, not many privileged patients use the facilities as their primary choice especially in big cities and urban areas, since they prefer private facilities.[45] Exceptions may occur when the cost of private care is unaffordable. It should be mentioned that some of the autonomous

---

[40] Ibidem.

[41] Dr. Prabhakar Kore Hospital and Medical Research Centre, Belgaum; Manipal Group of Hospitals; Dr. D. Y. Patil Medical College, Hospital and Research Centre Navi Mumbai; Christian Medical College, Vellore.

[42] King Edward Memorial (KEM) Hospital Mumbai; Rajawadi Hospital, Mumbai.

[43] Goa Medical College, Goa; Government Medical College Nagpur, Sir J. J. Hospital, Mumbai.

[44] A. Sengupta, S. Nundy, "The Private Health Sector in India."

[45] Ibidem. S. D. D'Cunha, "Despite a Booming Economy."

institutions[46] directly under the Government of India's Ministry of Health, do provide medical care that is world class and free or subsidised. However, such institutions are far and few, and beyond easy access for most of the population.

Having briefly described the Indian healthcare system, we will now endeavour to analyse how each of these sectors would have responded to a case like Charlie's or Alfie's.

C. Treatment in different setups

C.1. Private/Charitable/Trust/Missionary hospitals

Not many of these hospitals would have the resources to provide specialist paediatric intensive care facilities to treat complex cases such as those of Charlie's and Alfie's. Until recently, many hospitals run by charitable or religious organisations provided exemplary service but with increasing technology-driven medicine, they have fallen behind.[47] They are very likely to transfer the patient to a government-run facility for free care or enter into discussions with parents and relatives regarding the withdrawal of life-sustaining support in view of increasing costs which the family must bear (see infra).

C.2. Corporate hospitals

Although these hospitals have a good reputation regarding the quality of care provided, it is also said that in these hospitals, the "healthcare delivery has become a 'business' … corporate ethics are focused on profits rather than patient welfare."[48] For many corporate hospitals, "the patient is an asset for the first 48 hours because the patient will be subjected to most of the investigations and the earning is maximum. After 48 hours, the patient is a liability except when the patient is very sick and needs intensive care, another profitable area for the hospital."[49]

---

[46] All India Institute of Medical Sciences New Delhi; The Postgraduate Institute of Medical Education and Research (PGIMER) Chandigarh; Sree Chitra Tirunal Institute for Medical Sciences and Technology Thiruvananthapuram.
[47] A. Sengupta, S. Nundy, "The Private Health Sector in India."
[48] V. Gupta, M. Gupta, V. Gupta, "Violence against Doctors: Causes, Effects and Solutions," in *Progress in Medicine*, ed. Rohini Handa, vol. 28 (India: Evangel Publishing, 2018), 369.
[49] Ibidem, 369.

Therefore, without adequate financial resources, most ordinary Indians would struggle to cope with the costs of treating their child in an intensive care unit for a prolonged period. Corporate hospitals are not sworn by any ethical considerations and are even accused of artificially prolonging life by extraordinary means for financial gains,[50] refusing to admit patients until advance payments are made even during an emergency,[51] refusing to discharge patients[52] or release dead bodies until dues are paid.[53] So in this climate of financial greed and unethical practices, it is likely that the corporate hospitals would either try to persuade parents who are unable to afford the treatment to discontinue it and take the child home, or take the child to another hospital of their choice, viz. government-run hospital if available. In the case of disagreement with the patient or family, the treating team may even refuse treatment.[54] It is almost impossible to envisage a situation where, if the family can afford to pay the bills, the hospital would decline the parents' request for treatment continuation and approach the court for direction.

C.3. Government hospitals

Many government-run cottage and district general hospitals lack the expertise and resources to treat complex cases such as those of

---

[50] Ibidem, 367-372.

[51] "Hospital Keeps Dead Body Because Family Can't Pay Bill," *NDTV* (December 19, 2011). https://www.ndtv.com/mumbai-news/hospital-keeps-dead-body-because-family-cant-pay-bill-567201.

[52] "Detaining Patients for Unpaid Bills is Illegal: Bombay HC," *The Times of India* (January 12, 2018).

[53] "Hospital Keeps Dead Body Because Family Can't Pay Bill." S. Sukhi, "Hospital Refuses to Hand Over Body Over Dues," *The Times of India* (April 13, 2017). S. Basu, "Dead Bodies Can't Be Held 'Hostage' by Delhi Hospitals over Unpaid Bills, Says Draft Advisory," *NDTV* (May 29, 2018).

[54] "Code of Medical Ethics Regulations, 2002," *Medical Council of India*, https://www.mciindia.org/CMS/rules-regulations/code-of-medical-ethics-regulations-2002 (accessed July 17, 2018). M. Safi, "Doctors Treating World's Heaviest Woman Resign over Claims They Lied about Her Weight Loss," *The Guardian* (April 27, 2017), https://www.theguardian.com/world/2017/apr/27/doctors-treating-worlds-heaviest-woman-eman-ahmed-resign-claims-lied-weight-loss.

Charlie's or Alfie's. A few government teaching hospitals and central government-funded autonomous hospitals would, however, be in a position to manage such complex cases. Given that these hospitals are located either in the capital or in large cities, many rural patients find these facilities inaccessible due to distances and poor access to transportation. Thus, a very sick child from a rural area would almost certainly die of natural causes. If, however, cases like Charlie's or Alfie's were to be brought to these hospitals, would they continue the treatment after the prognosis became clear? These are some of the best known premier teaching hospitals practising evidence-based medicine. Purely based on medical evidence, it is unlikely that they would continue treating the patient for financial, religious, or any other reasons. But, what would happen and how would the hospital management react if the hospital came under political, public, and religious pressures, as happened in the UK? Would the hospital approach the court to seek permission to withdraw the life support under these circumstances? Since government-run institutions are dependent on government finances and open to political interference, including but not limited to, transfer of staff between rural and city hospitals,[55] it can be hypothesised that the hospital management would, due to the fear of retribution and apprehension of committing a criminal act (see the discussion on the Indian Legal System below), almost certainly continue with the expensive life-sustaining intensive care treatment such as ventilation against sound medical judgements.

D. The doctor-patient relationship in India

The doctor-patient relationship in India is significantly different than in the UK. Because of distinct socio-economic, cultural, and legal differences, as discussed below, a hospital management would be

---

[55] "Conspiracy behind SMO's Transfer: Docs," *The Times of India* (October 28, 2016), https://timesofindia.indiatimes.com/city/chandigarh/Conspiracy-behind-SMOs-transfer-Docs/articleshow/55105199.cms. "Chawla Cries Foul Over Transfer of Doctors," *Hindustan Times* (January 12, 2013), https://www.hindustantimes.com/punjab/chawla-cries-foul-over-transfer-of-doctors/story-Mz0oTBuJt2Uqz5Sg23HOzH.html. B. Purohit, T. Martineau, K. Sheikh, "Opening the Black Box of Transfer Systems in Public Sector Health Services in a Western State in India," *BMC Health Services Research* 16:1 (2016): 419.

reluctant to approach the court for discontinuation of treatment. In reality, the "law" seems to enter the picture only in case of negligence or malpractice.[56]

D.1. Multicultural society

India is still a deeply religious multicultural and multi-religious society. According to the last official census data in 2011, of the nearly 1.028 billion people, 80.5% identified themselves as Hindu while Muslims and Christian were the other two major religions representing 13.4%, and 2.3% respectively. Only 0.1% did not state their religion.[57] Over the centuries, these communities have lived in harmony and cross-cultural influences exist between different communities.

Although most Hindus respect life, they also believe that helping to end suffering in a painful life is akin to performing a good deed and part of moral obligation.[58] Islam believes life is provided by God and therefore killing is a sin.[59] Hindus and Muslims, based on their religious beliefs, are often more likely to accept the inevitable as God's wish in a terminal case, after they have exhausted all their resources. They are more agreeable to accept the discontinuation of intensive treatment, letting nature take its course unless they believed the treatment provided was incorrect. In such circumstances, relatives may approach a religious leader (*tantric*) or seek alternative (complementary) "pathies" during the final days.[60] On the other hand, the Catholic Church believes that life should be preserved from the moment of conception and seems to not accept that a painful or difficult or "non-productive" life

---

[56] Confirmed by Dr. M. Vaze via personal communication with the first author on September 8, 2018.

[57] Office of the Registrar General & Census Commissioner, India, "Religion," http://censusindia.gov.in/Census_And_You/religion.aspx (accessed September 5, 2018).

[58] BBC, "Hindu Attitudes Towards Euthanasia." http://www.bbc.co.uk/schools/gcsebitesize/rs/sanctity/hieuthanasiarev2.shtml- (accessed September 6, 2018).

[59] M. Healy, "The Sanctity of Life and World Religions," *National Catholic Register* (January 22, 2014).

[60] Conveyed by Dr. M. Vaze via personal communication with the first author on September 8, 2018.

is not worth living.[61] So, in a hospital run by a Christian missionary, one wonders whether the hospital management would favour the sanctity of life over medical prognosis and the patient's suffering. The authors can only surmise that in such an eventuality, the hospital may, unlike in the UK, continue the treatment or even agree to transfer a child to another facility such as the Vatican-linked Bambino Gesù Hospital in Rome which was willing to treat Charlie[62] and Alfie.[63]

D.2. Patient autonomy and doctor-patient communication

Traditionally, doctors in India have been highly respected and considered as next to God. Patients believe that doctors act in the best interests of a patient and would not do any harm. If an untoward incident happened, the patient would consider it to be God's will. Furthermore, doctors demonstrate "paternalistic" attitudes, which have led to doctors not taking patients into confidence, not asking them about their ideas, views and expectations, nor explaining to patients or relatives the reasons behind tests or treatment.[64] This is contrary to the UK practice where patient autonomy is crucial for the patient-centred consultation.[65] In the last few decades, due to increasing literacy, access to information, awareness,[66] and paying patients demanding better care, sometimes with unrealistic expectations,[67] the healthcare situation has been changing in India. However, in general, patient autonomy is neither recognised nor respected and the doctor-patient communication is poor.

D.3. Public opinion and violence against doctors

While the parents of Charlie and Alfie disagreed with their treating

---

[61] M. Healy, "The Sanctity of Life."

[62] S. D'Arcy, "Charlie Gard: Vatican Hospital Offers to Care for Terminally-ill Boy after Pope and Donald Trump Tweet Support," *Independent* (July 5, 2017).

[63] BBC, "Evans Parents Appeal," 2018.

[64] S. Paul, V. Bhatia, "Doctor Patient Relationship: Changing Scenario in India," *Asian Journal of Medical Sciences* 7:4 (2016): 1-5.

[65] GMC, "Good Medical Practice," 2013.

[66] V. Gupta et al, "Violence Against Doctors." S. Paul, V. Bhatia, "Doctor Patient Relationship."

[67] Ibidem (Gupta et al).

doctors, there was no evidence that they were unhappy with the quality of care provided by either of the hospitals. In fact, in his judgment in *Alder Hey Children's NHS Foundation Trust v Evans, James & Evans*, Mr. Justice Hayden said at paragraph 31 that "[t]he range of investigation that I have set out … has also to be considered in the context of the available facilities at this particular hospital, which can properly be said to *hold world-class facilities.*"[68] When the European Court of Human Rights (ECtHR) refused to intervene in the case, the supporters of Alfie demonstrated outside the Alder Hey Hospital and intimidated hospital staff[69] while some tried to storm the hospital,[70] although without physical violence against hospital staff or causing property damage. Although violence against NHS staff is not uncommon in the UK,[71] damage to property is far less common. Rare as it may be, offenders do face prosecution. In contrast to this, the Indian situation where dissatisfaction with the healthcare provided and/or any other reason related to care can often result in not only physical violence towards the staff,[72] but also in the destruction of the facilities.[73] Some of the other reasons said to have contributed to the recent spate of violence against doctors and hospitals are the poor staffing of government hospitals resulting in an often unmanageable doctor-patient ratio,[74] corruption within the healthcare system,[75] the

68 [2018] EWHC 308 (Fam) (our emphasis).

69 R. Hartley-Parkinson, "Alder Hey Hospital Staff are Receiving Threats and Personal Abuse from Alfie Evans' Supporters," *Metro* (April 26, 2018).

70 BBC News, "Alfie Evans: Protesters Try to Storm Alder Hey Hospital," (April 23, 2018).

71 R. Vize, "Rise in Assaults on Staff Reveals Intolerable Pressure on NHS," *The Guardian* (April 20, 2018).

72 "Patient's Kin Pull Out Guns, Beat Up Three Doctors at AIIMS Emergency," *The Indian Express* (July 15, 2014).

73 A. Gowen, "Angry Mobs Are Attacking Doctors at Hospitals in India," *The Washington Post* (May 24, 2016).

74 A. Sengupta, S. Nundy, "The Private Health Sector in India." A. Supe, "Violence against Doctors cannot be Tolerated," *British Medical Journal* (blog) (2017) https://blogs.bmj.com/bmj/2017/03/29/avinash-supe-violence-against-doctors-cannot-be-tolerated/ (accessed on May 5, 2019).

75 Ibidem (Sengupta & Nundy). V. Gupta et al, "Violence Against Doctors." S. Paul, V. Bhatia, "Doctor Patient Relationship." V. Mahajan, "White Coated Corruption," *Indian Journal of Medical Ethics* 7:1 (2010): 18-20.

misperception that doctors treat patients for financial gains,[76] the erosion of medical ethics,[77] and unethical practices.[78] Most doctors faced with violence are reluctant to register a complaint with the local police because of the perception that police are corrupt and easily bribed, as well as because of the protracted judicial process, an overall impression of a failure of various regulatory and law enforcing agencies, and the fear of creating enemies within the local community where they have to continue practising.[79] Approaching courts is not an easy option because violent riots against court decisions that people disagree with are not uncommon in India.[80] [81] One can thus hypothesise that the fear of violence and intimidation would be putting doctors off from going against the wishes of parents/relatives and not approach the court but rather continue treating the patient whilst evaluating other options such as mediation, coercion, or transfer to a government-run hospital.

D.4. Political interference

In the case of Charlie Gard, the US President Donald Trump, Vice-President Mike Pence, and many congressmen expressed their support for Charlie's parents.[82] Political interference in Indian healthcare system is not uncommon, especially when it comes to appointments and

---

[76] A. Supe, "Violence Against Doctors."

[77] V. Gupta et al, "Violence Against Doctors"; V. Mahajan, "White Coated Corruption."

[78] Ibidem (Mahajan).

[79] V. Gupta et al, "Violence Against doctors."

[80] "Deadly Dalit Caste Protests Hit India Spurred by Supreme Court Decision," *DW.COM* (April 3, 2018).

[81] "Twenty-eight Dead as Violence Erupts among Devotees of India's 'Guru of Bling' Following Rape Conviction," *The Telegraph* (August 25, 2017).

[82] B. Jacobs, H. Pidd, "Donald Trump Offers Help for Critically Ill Baby Charlie Gard," *The Guardian* (July 3, 2017). M. Henneberger, "What Mike Pence Gets Wrong - and Right - About Baby Charlie Gard and Health Care," *The Kansas City Star* (July 12, 2017). R. Revesz, "Charlie Gard: US Congressmen Want to Make Sick Baby an American Citizen so he can Receive Treatment Abroad," *Independent* (July 9, 2017).

transfers to different parts of the state or the country.[83] The patient or relatives often approach a politician in the hope of influencing medical management in providing care, seeking better facilities or reduction in charges and it is likely that the cause of a sick patient would be taken up by politicians for political gains.[84] It is postulated that many politicians interfere with the care and the cost provided to please their core voters.[85] In such circumstances, rather than facing the wrath of a politician,[86] hospital management would likely comply with such requests - "a medical professional finds it very difficult to refuse a VIP's request."[87]

D.5. Religious interference

In both Charlie's and Alfie's cases, Pope Francis offered prayers and expressed solidarity with the parents, even proposing to treat the children at the Vatican-owned Bambino Gesù Hospital in Rome.[88] Pro-life groups from the USA got involved in supporting Charlie's parents.[89] Although the Indian society is largely religious, the involvement of religious leaders in medical treatment is uncommon. Parents are more likely to carry out other religious activities such as making offerings at places of worship than approach a religious leader for intervention in hospital care.

---

[83] "Conspiracy Behind Transfer," *Times of India* (2016). "Chawla Cries Foul," *Hindustan Times*, 2013.

[84] S. Paul, V. Bhatia, "Doctor Patient Relationship."

[85] P. Ambesh, "Violence Against Doctors in the Indian Subcontinent: A Rising Bane," *Indian Heart Journal* 68:5 (2016): 749-750.

[86] N. Carvalho, "Catholic Hospital Attacked over a Land Dispute in Madhya Pradesh," *AsiaNews* (March 13, 2018). "Samajwadi Party Lawmaker Irfan Solanki Booked for Attack on Doctors," *NDTV* (March 7, 2014).

[87] V. Mahajan, "White Coated Corruption," 19.

[88] P. Greenfield, "Charlie Gard: Pope Shows Solidarity with Parents of Critically Ill 10-month-old," *The Guardian* (July 3, 2017). C. Collins, "Pope Francis Asks that Alfie Evans be Allowed to 'Seek New Forms of Treatment'," *Crux* (April 23, 2018).

[89] J. Glenza, "How Charlie Gard Captured Trump's Attention and Animated Pro-Life Groups," *The Guardian* (July 26, 2017). J. Halliday, "'Call from God': American Pro-Lifer's Role in Alfie Evans' Battle," *The Guardian* (28 April 2018).

D.6. Socio-economic framework

The parents of both Charlie and Alfie were not financially privileged. Charlie's mother Connie was a carer and his father Chris worked as a postman.[90] Whilst Charlie was treated for nearly nine months (September 2016 – July 2017)[91] at GOSH, Alfie was hospitalised at Alder Hey for nearly 16 months (December 2016 – April 2018).[92] No Indian family from a similar working class background could afford private intensive care for such a prolonged period of time because, as Sengupta and Nundy explain, "more than 40% of all patients admitted to hospital have to borrow money or sell assets, including inherited property and farmland, to cover expenses."[93] According to a neonatologist consultant from Mumbai,[94] the cost in a paediatric intensive care unit in a corporate hospital may vary between INR 10,000 – 20,000 (USD 136-272)[95] per day, whereas the average gross annual income of an Indian graduate is INR 1,191,445;[96] unskilled and semiskilled earn much less. Although the parents of Charlie and Alfie did not pay for their NHS treatment, they and their supporters raised money through crowdfunding[97] for overseas treatment. They achieved this with the help of social media. In India too, there are charitable as well as non-government organisations (NGOs) and at least in one case a charity approached the High Court compelling a government hospital to provide expensive treatment for a child with a rare genetic disorder.[98]

---

[90] R. Wheatstone, S. Petkar, "Brave Mum and Dad who are Connie Yates and Chris Gard?," *The Sun* (June 2, 2018).

[91] G. Heffer, "Timeline: Parents' Battle to Save Charlie Gard," *Sky News*, https://news.sky.com/story/timeline-parents-battle-to-save-charlie-gard-10914755 (accessed August 27, 2018).

[92] "Timeline: Key Events in the Legal Battle over Alfie Evans," *The Guardian* (April 28, 2018).

[93] A. Sengupta, S. Nundy, "The Private Health Sector in India."

[94] Dr. A. Gajendragadkar.

[95] Via personal communication with the first author.

[96] "What is Average Salary in India?" *Average Salary Survey.* https://www.averagesalarysurvey.com/india (accessed October 18, 2018).

[97] "Well-wishers Raise More Than £65,000 to get Treatment for Alfie Evans." *The Catholic Universe* (February 22, 2018).

[98] A. Kazmin, "Indian Parents Fight in Court for Access to Costly Medical Care," *Financial Times* (28 February 2017).

Therefore, if a well-equipped government-run hospital was not accessible, could an Indian family raise funds to treat their child in an expensive corporate hospital? There is evidence that Indians have donated generously through crowdfunding.[99] Nonetheless, these were isolated contributions and it is anyone's guess whether crowdfunding could raise adequate funds to sustain complex intensive care treatment for several months in a corporate hospital. We would not like to speculate whether the recently-introduced universal healthcare system[100] would pay for expensive treatments in private/corporate hospitals or prefer to transfer the patient to a government-run facility to reduce costs.

Having discussed the differences in the medical structure and facilities between the two countries, we now consider how Indian courts could have reacted.

III. The Indian legal system

After gaining independence in 1947, India became a democratic republic and its Constitution came into force on 26th January 1950. Its independent judiciary includes the Supreme Court of India (SCI) as the highest appellate court. Each federal state has a High Court with subordinate civil and criminal courts. The Law of the Union is superior to that of the States.[101] Although the Indian legal system has many similarities with the English legal system (on which it is based, due to the colonial rule, incorporating many laws from the colonial era), "the Indian system incorporates the common law system along with the

[99] "Social Media Campaign Raises Small Fortune for Family of Indian Man Killed Cleaning Sewer," *ABC News,* http://www.abc.net.au/news/2018-09-20/money-raised-for-india-family-after-photo-of-boy-goes-viral/10284660 (accessed September 27, 2018). "Crowdfunding Pays Hospital Bills of Injured Indian Girl," *BBC News*, https://www.bbc.co.uk/news/world-asia-india-42942653 (accessed September 27, 2018).

[100] A. Withnall, "Modicare: India Launches World's Biggest Experiment."

[101] "Indian Law: Legal system," *Oxford Lib Guides. Bodleian Libraries*, https://ox.libguides.com/c.php?g=422964&p=2888488 (accessed September 15, 2018).

statutory law and the regulatory law."[102] In addition, the High Courts and the Supreme Court have the power of judicial review to invalidate the actions of the legislative as well as executives "if they are *ultra vires* of the Constitutional provisions."[103]

Legal considerations

In Charlie's and Alfie's cases, the English courts and the ECtHR had to decide whether medical treatment could be withdrawn/withheld against parental wishes when medical opinion concluded that the infants' conditions were progressive and terminal, and that prolonging the treatment was not in their best interests. Thus, the patients' interests and wellbeing (suffering), supported by medical evidence, was given more relevance (importance) than the wishes of the parents. This is because in England and Wales, disputes between parents and doctors must be referred to the courts. Judges will decide, based on medical evidence alone, as shown in those two cases, what is in the child's best interests.

In India, on the contrary, until recently there were no clear regulations or legal principles regarding the withdrawal/withholding of medical treatment from patients. The provision of brain death declaration did not exist until India enacted the Transplantation of Human Organs Act 1994.[104] Even then, uncertainty persisted whether "futile" treatment could be withdrawn/withheld in cases where the prognosis of the patient was considered poor but the patient was not "brain dead." Another area of confusion within the medical and legal professions concerned what action may be perceived as "abetting suicide" or "culpable homicide." This is because in India, unlike in England and Wales, attempting suicide is a penal crime under the Indian Penal Code (IPC) section 309 and is dealt with simple imprisonment up to one year and/or a fine.[105] Furthermore, any

---

102 "SILF - Indian Judicial System," *Society of Indian Law Firms*, http://www.silf.org.in/indian-judicial-system.php (accessed September 15, 2018).

103 Ibidem.

104 "The Transplantation of Human Organs Act, 1994," *Central Government Act*, https://indiankanoon.org/doc/1666741/ (accessed September 15, 2018).

105 "The Indian Penal Code," *Central Government Act*,

individual who "abets the commission of such suicide" may be punished with imprisonment of up to ten years and/or be fined under section 306 of IPC.[106] Doctors therefore feared that withdrawing life-sustaining treatment could be viewed as abetting suicide and therefore often shied away from withdrawal or withholding of treatment.[107] The 2002 regulations of the Medical Council of India classed euthanasia as "unethical conduct" without specifically addressing the issue of withholding or withdrawing care in terminally ill patients or patients with poor prognoses.[108] The belief that withdrawing and withholding care may be classified as "passive euthanasia" leading to litigation persisted.[109] In fact, the then Union Law Minister of India is said to have opined that "we are of the view that euthanasia and assisted suicide must continue to be offences under our law" and the legal profession in India seems to be confusing withdrawal/withholding of care with passive euthanasia.[110] This explains why many doctors and hospitals continue treating patients in expensive intensive care units even when the prognoses are poor. This is despite the fact that patients or relatives often approach doctors requesting them to limit or withdraw treatment either because the futility of the treatment becomes evident after a few days or because of mounting financial burden.[111] If there is a disagreement (with the doctors), families take the patient away (from a hospital) against medical advice i.e. leave against medical advice (LAMA)[112] rather than pursue legal avenues. In the UK, on the contrary, patients who discharge themselves against medical advice (DAMA) seem to be doing so because of the distress caused by living

https://indiankanoon.org/doc/1569253/ (accessed September 5, 2018).

106 Ibidem.

107 M. Yadav, "End of Life Care Support: Ethical and Legal Scenario in India," *Journal of Indian Academy of Forensic Medicine* 28:3 (2006): 93-97.

108 Indian Medical Council (Professional Conduct, Etiquette and Ethics) Regulations (2002), https://www.mciindia.org/documents/rulesAndRegulations/Ethics%20Regulations-2002.pdf (accessed September 16, 2018).

109 M. Yadav, "End of Life Care Support."

110 Ibidem.

111 Ibidem.

112 Ibidem.

in hospital rather than poor care or financial difficulties *per se*.[113]

The issue of passive euthanasia came to the fore when the SCI was confronted with a case relating to persistent vegetative state (PVS). King Edward Memorial (KEM) Hospital in Mumbai, a well-known teaching institute, was approached by one Ms. Pinki Virani of Mumbai, claiming to be a next friend, to withdraw/withhold feeding from Aruna Ramchandra Shanbaug who had then been in PVS for nearly 37 years. The court appointed a panel of three medical experts who confirmed that the patient was indeed in a persistent vegetative state. The hospital's (primarily nursing) staff, whom the court recognised as "the family," had opposed withdrawal of feeding as did the state and the Union Governments. In 2011, the apex court delivered its judgment[114] and denied the request, dismissing the case. In their judgment, Markandey Katju J and Gyan Sudha Mishra J emphasised that:

> A decision has to be taken to discontinue life support either by the parents or the spouse or other close relatives, or in the absence of any of them, such a decision can be taken even by a person or a body of persons acting as a next friend. It can also be taken by the doctors attending the patient. However, the decision should be taken *bona fide* in the best interests of the patient.[115]

The court thus recognised the rights of parents/relatives/friends in the decision–making process. In paragraph 127, the judges observed that:

> considering the low ethical levels prevailing in our society today and the rampant commercialisation and corruption, we cannot rule out the possibility that unscrupulous persons with the help of some unscrupulous doctors may fabricate material to show that it is a terminal case with no chance of recovery.

---

[113] L. L. Machin, D. Goodwin, D. Warriner, "An Alternative View of Self-Discharge Against Medical Advice: An Opportunity to Demonstrate Empathy, Empowerment, and Care," *Qualitative Health Research,* 28:5 (2008): 702–710. doi: 10.1177/1049732318754514.

[114] *Aruna Ramchandra Shanbaug v Union of India & Ors* AIR 2011 SC 1290.

[115] Ibidem, para. 126(i).

The court then concluded:

> In our opinion, in the case of an incompetent person who is unable to take a decision whether to withdraw life support or not, it is the court alone, as *parens patriae*, which ultimately must take this decision, though, no doubt, the views of the near relatives, next friend and doctors must be given due weight.[116]

This case laid down the process to be followed by High Courts whenever either doctors or relatives wish to withdraw the treatment. This judgment differs from the English case of *Airedale NHS Trust v Bland*.[117]

In the same judgment, the SCI defined that such an act would be considered passive euthanasia:

> Passive euthanasia is usually defined as withdrawing medical treatment with a deliberate intention of causing the patient's death. For example, if a patient requires kidney dialysis to survive, not giving dialysis although the machine is available, is passive euthanasia. Similarly, if a patient is in a coma or on a heart lung machine, withdrawing of the machine will ordinarily result in passive euthanasia. Similarly, not giving lifesaving medicines like antibiotics in certain situations may result in passive euthanasia. Denying food to a person in a coma or PVS may also amount to passive euthanasia.[118]

This is not how passive euthanasia is interpreted by English Law.

The Law Commission of India, in its report no. 241,[119] considered the issues raised by the Aruna Shanbaug judgment, while in 2012, the Indian Society of Critical Care Medicine published the consensus Ethical Position Statement for end-of-life and palliative care in Indian

---

116 Ibidem, para. 132.

117 (1993) All E.R. (82) (H.L.)

118 *Aruna Ramchandra Shanbaug* (n 114), para. 51.

119 "Passive Euthanasia – A Relook," *Law Commission of India Report*, no. 241 (August 2012), http://lawcommissionofindia.nic.in/reports/report241.pdf (accessed September 10, 2018).

intensive care units.[120] Yet they failed to provide clear guidance regarding the withdrawal/withholding of treatment in terminally ill patients. In the absence of clear guidance, there still appeared to be confusion and apprehension within the medical fraternity about murder charges doctor(s) may face for withholding/withdrawing life-sustaining treatment, such as ventilator support, even if this has been requested by the patient's relatives.[121] In fact, in the four years after the Supreme Court's decision in the Aruna Shanbaug case, no one had appealed to the court for this action.[122] This demonstrates that even when doctors and the family are in agreement regarding the withdrawal of life support from terminally ill patients,[123] doctors are worried about seeking the opinion of medical experts and involving courts in the decision-making. It can therefore be surmised that most Indian hospitals would not have approached the courts for direction regarding the withdrawal of life-sustaining treatment from child patients such as Charlie or Alfie. Even if any hospital were to approach the courts, the likelihood is that the courts would have sided with the parents, just as did the SCI.

The second question that needs addressing is whether legal aid would be available to parents if they wished to challenge the medical opinion in the courts. The short answer is "yes." It is also possible that an NGO could support the parents (see *supra*). The courts would be guided by the SCI judgment.

But things could be changing. In March 2018, a five-judge constitutional branch of the SCI delivered its verdict in the case of a

---

[120] R. K. Mani et al, "Guidelines for End-of-Life and Palliative Care in Indian Intensive Care Units: ISCCM Consensus Ethical Position Statement," *Indian Journal of Critical Care Medicine* 16:3 (2012): 166–181. doi: 10.4103/0972-5229.102112.

[121] "Forcing Ventilator on Dying Patients is Assault," *The Times of India* (July 20, 2014). "Mumbai: In Debate over Life Support, Can't Afford Bill, Says Patient's Family," *The Indian Express* (April 13, 2017). R. Gursahani, R. Kumar Mani, "India: Not a Country to Die In," *Indian Journal of Medical Ethics* 1:1 (2016): 30-35.

[122] G. Mudur, 'Doctors Raise Concerns About India's Draft Bill to Regulate Life Support Withdrawal," *British Medical Journal* 353 (2016): i2919. R. Nagarajan, "Between Life Support and Death," *The Times of India* (May 24, 2015).

[123] Ibidem (Mudur).

public interest litigation (PIL) writ filed by Common Cause, a registered society, against the Union of India.[124] The case addressed many of the issues discussed above. In disposing of the writ, Dipak Misra CJI (Chief Justice of India) and A. M. Khanwilkar J in their seriatim judgment said:

> A competent person who has come of age has the right to refuse specific treatment or all treatment or opt for an alternative treatment, even if such decision entails a risk of death ... where a patient has already made a valid advance directive which is free from reasonable doubt and specifying that he/she does not wish to be treated, then such directive has to be given effect to.[125]

They further added that: "[w]hen passive euthanasia as a situational palliative measure becomes applicable, the best interests of the patient shall override the State interest."[126] In his concurring judgement, Dr. D. Y. Chandrachud J argued that "the treating doctor who, in a good faith exercise of professional medical judgment abides by an advance directive is protected against the burden of criminal liability"[127] and added that:

> The decision by a treating doctor to withhold or withdraw medical intervention in the case of a patient in the terminal stage of illness or in a persistently vegetative state or the like where artificial intervention will merely prolong the suffering and agony of the patient is protected by the law. Where the doctor has acted in such a case in the best interests of the patient and in *bona fide* discharge of the duty of care, the law will protect the reasonable exercise of a professional decision.[128]

Mr. Ashok Bhushan J, in his separate judgement, said that the SCI had already declared that the right to die with dignity was a constitutional right. In his conclusions, he stated:

---

124 *Common Cause (A Regd. Society) v Union of India* AIR 2018 SC 1665.

125 Ibidem, para. 195(viii).

126 Ibidem, para. 195(xiv).

127 Ibidem, para. 143(vi).

128 Ibidem, para. 143(vii).

> We are thus of the opinion that the right not to take a life-saving treatment by a person, who is competent to take an informed decision is not covered by the concept of euthanasia as it is commonly understood but a decision to withdraw life-saving treatment by a patient who is competent to take decision as well as with regard to a patient who is not competent to take decision can be termed as passive euthanasia, which is lawful and legally permissible in this country.[129]

He further added that:

> We also are of the opinion that in cases of incompetent patients who are unable to take an informed decision, "the best interests principle" be applied and such decision be taken by specified competent medical experts and be implemented after providing a cooling period to enable aggrieved persons to approach the court of law.[130]

The judgment thus clarifies that the Indian Constitution acknowledges the individual's constitutional right to die with dignity, that abiding by a competent person's advance directive to refuse treatment would not imply abetting suicide or culpable homicide and that doctors and/or family members could seek directives from the High Court regarding the withholding/withdrawal of treatment in terminal cases. The judgment should assure doctors that they would not face legal repercussions if they acted in the best interests of their patients.

In view of the above, if a case like Charlie or Alfie were to occur in India and if the case was to reach the court prior to the March 2018 SCI judgment, it is likely that the court would have sided with the parents as it did in the case of Aruna Shanbaug because the infants were neither "brain dead" nor in PVS. The March 2018 judgment from the SCI makes it possible for doctors to be more realistic and be guided by the medical evidence in such challenging situations. However, given that the SCI has dealt, so far, only with adults, we cannot speculate how it would decide if it were to consider the case of a child with a

[129] Ibidem.

[130] Ibidem, para. 143(ix).

deteriorating condition and poor prognosis. Until then, we believe that Indian doctors should be assured that they will not be prosecuted for committing passive euthanasia or abetting suicide if, when withholding/withdrawing life-sustaining care, they followed the prescribed procedures and acted in the best interests of their patients based on medical evidence. Nevertheless, we believe that doctors and parents/relatives should make every effort and take all possible steps to reach an amicable decision that best serves the interests of the patient.

IV. Summary

The cases of Charlie Gard and Alfie Evans are tragic for many reasons. We accept that both of them received the best medical care possible and that they would have had the same (terminal) outcomes irrespective of the country they were treated in.

Although India is a developing country, the health services are poorly organised and the government spends only a small percentage of the GDP on healthcare compared to many other developed/developing countries. Access to medical care in general, and specialised care in particular, is restricted not only by geographical location but also by financial affordability. Most government-funded hospitals, unlike the NHS, are poorly resourced and patients often pay out of their own pockets for private healthcare. Quality healthcare provision does not seem to be high on the political agenda. Private care is expensive and prolonged intensive care in a private hospital would burden most families. The concepts of "death with dignity" and "withholding/withdrawing of treatment" are not well understood by the society. Until recently, the medical profession was understandably apprehensive and worried about the legal ramifications of withholding/withdrawal of treatment because of the absence of clear legal guidance, confusion regarding passive euthanasia and lack of unambiguous advice from professional bodies. This is compounded by the fact that people seem to be losing their trust in the medical profession and view doctors as greedy, corrupt, uncaring, and not always putting their patients' interests first; an image partly created by media reports. The Supreme Court of India, in its two judgments of 2011 and 2018, has clarified the legal position regarding several of these issues and laid down procedures for doctors and individuals to follow. These guidelines should reassure patients, families, and doctors that acting in the best interests of their patient based solely on medical evidence will not attract legal ramifications.

Therefore, in India, until recently, a child such as Charlie or Alfie would most likely have died sooner of natural causes either because of lack of access to specialist medical services or due to an inability to fund the treatment in a private intensive care unit. If, however, treatment was commenced, it could not be withdrawn/withheld for fear of being criminally charged for abetting suicide/manslaughter – unlike in England and Wales. Similarly, approaching the High Court and/or the Supreme Court would not only be expensive but would have faced the snail pace challenge of the Indian legal system.

However, after the SCI judgment of 2018, it is conceivable that doctors treating a case similar to Charlie's or Alfie's would feel much more comfortable to act in the best interests of their patient, guided by the medical evidence alone, and seek the courts' direction to that effect if necessary. The sad thing is, irrespective of what the courts would have decided, it would not have changed the fate of those patients.

Chapter 15

Medical Futility and Parental Paternalism in Turkey

Banu Buruk,[1] Berna Arda[2]
[1] Scientific and Technological Research Council, Turkey
[2] Ankara University, Turkey

I. Introduction

Medical futility is a sensitive issue in paediatrics. The parties involved in the treatment process – both the parents and the physician – share the same wish: "the best interests of the child." However, parents may sometimes choose different means of achieving this from the physician. When this happens, the physician's decision-making process becomes more complicated as he or she has to respect the parents' autonomy while also protecting the child's well-being. In such situations, disagreements between the physician and the parents are likely to occur. The disagreement may be attributable to the parents' lack of information or misinformation, different value systems, sentimentalism, and their social environment. The parents' social environment may differ in communitarian "we societies" versus individualistic "I societies." The Turkish social environment, which is generally accepted as a "we society," is usually formed of their relatives. Therefore, a disagreement between the physician and the parents may transform into a disagreement between the physician and the relatives of the parents.

Turkey combines western and eastern values. However, the large number of refugees from Syria provides Turkey's population with more diversity in terms of cultures. In this chapter, we will first briefly describe the terms of futility, autonomy, and paternalism. We will then interpret them within the Turkish medico-legal context. We will explore the conditions relating to disagreements between physician and parents, by giving some refugee statistics that are unique to Turkey. We will conclude by reflecting on how situations like Charlie Gard and Alfie Evans would be dealt with in Turkish society.

II. Futility, autonomy, and paternalism

According to the experiences of physicians and their consideration of empirical data in the scientific literature, if a medical treatment was useless in the last 100 cases, then that treatment ought to be declared futile.[1] For the most part, medical futility is understood as a medical treatment which lacks beneficial results, such as extending life span, reducing pain, and enabling comfort. The probability that the same treatment, in the 101st case, is futile will be very high. However, this remains a probability and will never be 100% certain. Although the treatment is the same in all cases, each case is unique; none will ever be identical. This uncertainty keeps parents' hope high, especially when they are seeking beneficial treatment for their child. For rare diseases, the calculation of the futility rate of a particular treatment is very difficult, sometimes impossible. Thus if a child is suffering from a rare disease, every attempt of any treatment will give hope to parents. In medical cases, being futile is expressed for treatments which are not hopeless, but are empirically inestimable.[1] Hopelessness is the feeling that no treatment will be beneficial. However, medical futility is a complex and subjective situation, and a value-dependent concept which contains some degree of uncertainty.[2] Although both physicians and parents have the best interests of the child at the forefront of their minds, the uncertainty of a futile treatment may cause disagreements between them. The case of Charlie Gard is an example of such a disagreement on futile treatment. Diagnosed with a rare severe mitochondrial disorder, his doctors and the courts agreed that continuing medical treatment was not in Charlie's best interests. Still, much of the media attention in Charlie's case was focused on the rights of the parents, who disagreed with the doctors, in the decision-making process.[3] A similar case was that of Alfie Evans who suffered from a severe neurodegenerative disorder. Following 16 months of continued intensive care, Alfie's doctors decided that intensive care was futile. The courts agreed and

---

[1] L. J. Schneiderman et al, "Medical Futility: Its Meaning and Ethical Implications," *Annals of Internal Medicine* 112 (1990): 949-954.
[2] M. Aghabarary, N. D. Nayeri, "Medical Futility and Its Challenges: A Review Study," *Journal of Medical Ethics and History of Medicine* 9 (2016): 11.
[3] D. Wilkinson, J. Savulescu, "Hard Lessons: Learning from the Charlie Gard Case," *Journal of Medical Ethics* (2017): 1-5.

similarly ruled that life-sustaining treatment was not in Alfie's best interests. However, his parents wanted to continue with life support.[4] Both these cases were traumatic for the medical professionals and the families.

The concept of futility itself has always been a source of debate. The main contentions were in relation to the goals, effect, the utility of the treatment in question, and values. The goals of the treatment depend on the effect and the utility components. If the treatment is effective and the probability of its direct and indirect benefits is high, then the futility degree of that treatment seems to be low. However, in general, the concept of futility is individualistic and based on the unique conditions of each patient.[5] The subjective side of medical futility raises some concerns for the patient and his or her family members/parents, such as: "what if a valuable treatment is discontinued because the treatment is considered futile from a medical perspective?" or "what if the essential treatment for the child is labelled as futile in order to cut healthcare costs?" As a result, the pronouncement of futility is difficult in medical cases. Moreover, the uncertainty of medical futility is also dependent on the question of "who will describe which treatment is futile and which one is not?" This question is directly related to the issue of authority over the continuation or discontinuation of futile treatment. Accordingly, it makes the decision-making process problematic when the decisions of the physician and the patient/parents do not match. In such cases, the physician's authority and the parents' autonomy stand in opposition to each other, and disagreement on the health status of the patient/child becomes inevitable.

The health status of every person is unique and important. Having a good level of health is a basic right. According to Article 25 of the Universal Declaration of Human Rights, "everyone has the right to a standard of living adequate for the health and well-being of himself and of his family, ... in circumstances beyond his control" (the Declaration of Human Rights, 1948). Here, the syntagm "beyond his control" is important from a medico-ethical perspective as respect for autonomy is one of the main ethical principles in medicine. Autonomy is self-

[4] D. Wilkinson, J. Savulescu, "Alfie Evans and Charlie Gard—Should the Law Change?" *Journal of Medical Ethics* 361 (2018): Editorial.

[5] M. Aghabarary, N. D. Nayeri, "Medical Futility and Its Challenges."

determination. Autonomy allows people to make decisions by themselves and for themselves without any controlling interference by others and without limitations, such as inadequate understanding. Including free will and adequate information, autonomy is a right to act rationally even if in the end the results are not satisfying. There is nevertheless the satisfaction of knowing that the mistakes would be one's own.[6] The concept of autonomy is undoubtedly important in 'we societies'; however, this concept should always be balanced with the concept of dependence on the community. It is clear that there is a direct correlation between autonomy and rational persons. The ability of rational decision-making needs 4 standards: (1) expressing a choice; (2) understanding; (3) reasoning; (4) appreciation. If a person cannot meet any of these 4 standards, then that person cannot be considered a rational human being.[7] Non-rational human beings such as children are vulnerable; therefore they need protection (mostly parental).

In "we societies," parental protection generally occurs in a paternalistic manner. However, a high level of parental paternalism results in the formation of a societal system and values based solely on adults.[8] Although Turkey combines western and eastern worldviews, eastern rules are generally given more weight. In the eastern world, or in "we societies," society as a whole is more important than individuals. In "we societies," others have the right to protect the best interests of a single person, because they may know best. Here "others" refer to a person or a group of people authorized to act in this capacity according to that society's cultural, legal, and moral norms. These authorised persons may be the leader of a society, the oldest male among the relatives, the father of the child, or the physician. Consequently, in "we societies," the adult-dominated approach is effective when refusing a crucial treatment or when insisting on a futile treatment.

---

[6] R. Munson, "Autonomy, Paternalism," in *The Interventions and Reflections; Basic Issues in Bioethics Book*, ed. Joann Kozyrev (Boston: Wadswoth, 2012), 38-50.

[7] P. Grotens-Wiegers et al, "Medical Decision-making in Children and Adolescents: Developmental and Neuroscientific Aspects," *BMC Pediatrics* 17 (2017) 120-130.

[8] C. H. Guvercin, B. Arda, "Parents Refusing Treatment of the Child: A Discussion About Child's Health Right and Parental Paternalism," *Clinical Ethics* 8 (2013): 52–60.

III. Current state and regulations in Turkey

In 2017, Turkey's population was over 80 million. Almost one quarter of the total population was under the age of 14 (Table. 1). For female and male individuals, the shared life expectancy at birth was 78 in 2017.[9] National health services are canalised to city hospitals, five of them having been put into service until today. City hospitals are big hospital complexes combining all the main hospitals in the city. The orientation of city hospitals is a result of the health transformative program. It is planned that 28 city hospitals will be put into service until 2021. As a result, by the year 2021, Turkey should have 41,000 extra bed spaces for patients.[10] In addition family practices instead of health centers are being promoted in the last few years. Inevitably, therapeutic medicine gained primary importance and preventive medicine was relegated to secondary importance. As a result of therapeutic medicine and new technologies and drugs, public health expenses multiplied fourfold between 2002 and 2014.[11]

Table 1. The population of Turkey by years and age groups, 2016-2017

Years	Population census		
	Age group	***Number of people***	*%*
2016	Total	79,814,871	100
2016	0-14	18,925,782	23.7
	15-64	54,237,586	68.0
	65 +	6,651,503	8.3

[9] Turkish Statistical Institute, *Statistics on the Number of Physicians in Turkey,* http://www.tuik.gov.tr/Start.do (accessed October 13, 2018).

[10] Turkish Ministry of Health, *Statistics on the Number of City Hospital's Bed Space,* http://www.saglikyatirimlari.gov.tr/TR,33960/sehir-hastaneleri.html (accessed October 30, 2018).

[11] Y. Tokatlioglu, I. Tokatlioglu, "Catastrophic Health Expenditures in Turkey and the Determinants of These Expenditures: 2002-2014," *Sosyoekonomi* 26 (2018): 59-78 (in Turkish).

Years	Population census		
	Age group	***Number of people***	**%**
2017	Total	80,810,525	100
	0-14	19,033,488	23.6
	15-64	54,881,652	67.9
	65 +	6,895,385	8.5
Source: Turkish Statistical Institute (TSI)			

In Turkey, the number of health professionals per person was 551 in 2016. However, the number of patient hospital visits per physician in the same year was 4,735 (Table. 2). Additionally, in 2016, the total health expenditure was 119,756,000 Turkish Liras and the total number of medical institutions was 32,980 (including community health centers, family medicine units, tuberculosis dispensaries, early cancer diagnosis, screening and training centers, specialty medical centers, and private outpatients clinics).[12] Keeping in mind that minors constitute a meaningful proportion of the country's total population, the number of physicians per child patients, the number of medical institutions dealing with minors, and the health expenditure spent for child patients also constitute a meaningful ratio of all these statistics.

Table 2. The number of physicians, number of persons per physician and number of patient hospital visits per physician in Turkey, 2009-2016

Years	Population census		
	Number of physicians	***Number of persons per physicians***	***Number of patient hospital visits per physician***
2009	118,641	612	4,447

[12] Turkish Statistical Institute, *Statistics on the Number of Physicians in Turkey.*

Years	Population census		
	Number of physicians	***Number of persons per physicians***	***Number of patient hospital visits per physician***
2010	123,447	597	4,367
2011	126,029	593	4,850
2012	129,772	583	4,791
2013	133,775	573	4,712
2014	135,616	573	4,749
2015	141,259	557	4,673
2016	144,827	551	4,735
Source: Turkish Statistical Institute (TSI) All physicians working in public and private sectors are covered.			

Turkey approves the main international agreements mentioning the protection of children and the rights of child patients. According to Article 90 of the Constitution of the Republic of Turkey, the international agreements that are approved by Turkey predominate over the national legislations.[13] Article 17 of the Constitution states that "the corporeal integrity of the individual shall not be violated except under medical necessity and in cases prescribed by law; and shall not be subjected to scientific or medical experiments without his/her consent." This Article is effective to prevent medical paternalism and allows adult patients or the legal guardians of children under 18 to be more autonomous in cases of medical futility.

Turkey signed the Convention on the Rights of the Child in 1990 and ratified it in 1994. The Convention entered into force as Law no.

[13] *Constitution of the Republic of Turkey,* Law No. 2709. Acceptance date November 7, 1982, https://global.tbmm.gov.tr/docs/constitution_en.pdf (accessed February 16, 2019).

4058 in 1995. Firstly, according to Article 28 of the Turkish Civil Code, personhood starts at the birth of the child in a healthy condition. According to the Law, personhood situation needs a healthy birth; the Article does not touch on the personhood situation of inborn disabled children. However, the capacity of having rights is already gained from the time of conception.[14] In addition, the term "child" refers to anyone below 18 years of age according to Article 6 of the Turkish Criminal Law (TCL).[15] Furthermore, according to Article 90 of the Turkish Criminal Law, scientific experiments on children are permitted only if the indispensable conditions below are met (TCL):

- It is essential that the scientific knowledge gained at the end of the study be dependent on the results of the experiments conducted on children;
- The written consent of the parents or the legal guardians should be obtained, as well as the verbal consent of the child who is capable of expressing his/her choices;
- There must be at least one paediatrician on the boards authorised to approve the study.

As seen from these two national legislations, starting from conception to the age of 18, children have to be protected by both their parents and paediatricians in medical treatments or scientific experiments. However, Article 24 of the Turkish Regulation of Patients Right which came into force in 1998 says that the responsibility for ensuring the health status of the child should belong to the physician during vital emergency conditions. In these kinds of situations, the consent of the parents/legal guardians is not an obligatory matter. In addition, if there is a need to perform a medical intervention on the child and the parents/legal guardians are not giving their consent for it; then the decision of the court takes precedence. In case of a disagreement between the physician and the parents/legal guardians, the physician can perform a medical intervention on the child with a supportive court decision.

---

[14] *Turkish Civil Code,* Law No. 4721. Acceptance date November 22, 2011, https://www.tbmm.gov.tr/kanunlar/k4721.html (accessed October 13, 2018).
[15] *Turkish Criminal Law,* Law No. 5237. Acceptance date September 26, 2004, https://www.tbmm.gov.tr/kanunlar/k5237.html (accessed October 13, 2018).

The Turkish Law for the Protection of Children, which entered into force in 2005, emphasises the principle of the "high benefit of the child" especially in Articles 5 and 45. These Articles mention that the law firstly confers the responsibility of ensuring the well-being of the child on the parents or legal guardians. However, if there is a need to take precautions for the protection of the child and if these precautions are related with the health status of the child, then the Ministry of Health takes the responsibility for protecting the physical and mental well-being of the child. The law points out that in case of a temporary or permanent treatment and/or rehabilitation of the child in question, the physician is in charge even if there is going to be a disagreement between the physician and the parents or legal guardians of the child.[16]

IV. Disagreements between physician and parents

Medicine is a form of applied beneficence and so is parenting. In terms of ethical principles, parenting and the medical profession share the same value of beneficence. Beneficence is actively doing good, helping others further their interests.[17] Beneficence also includes preventing harm, also known in bioethics as the principle of non-maleficence. In fact, the principle of beneficence consists of different components due to the timing of doing active goodness. According to the biomedical ethics principle of "beneficence;" first, before any harm may occur, we should prevent it, thus keeping stable the status of beneficence. Second, if any harm occurs, we should remove it and return to the status of beneficence. Third, in order to prevent harm, we should promote active goodness, which implies that we should always apply beneficence. Parents promote goodness actively for their children, always prevent harm and immediately remove it whenever it occurs; thus the physicians taking care of children should act in the same way. Moreover, parents and physicians promote beneficence for children to further the latter's interests, before they ask for help or before they notice anything going wrong with their health status. This situation forms somewhat protec-

---

[16] *Law for the Protection of Children,* Law No. 5395. Acceptance date July 3, 2005. http://www.mevzuat.gov.tr/MevzuatMetin/1.5.5395.pdf (accessed October 13, 2018).

[17] T. L. Beauchamp, J. F. Childress, *Principles of Biomedical Ethics* (New York: Oxford University Press, 2012), 202-241.

tionist and paternalistic parents and physicians. When the level of protectionism increases, the probability of disagreements between parents and physicians also increases. The main disagreement between physicians and parents is grounded in the conflict between their different views on what is most beneficial to the child. Parents act deeply dependent on their moral values; this is a natural instinct which cannot be undervalued. Similarly, physicians make decisions on an individual level and this level of decision-making always involves personal value judgments. However, the level of medical information they have allows physicians to stand by their observations. When the parents' decision-making process becomes less the product of observation analysis and reasoning, they become more optimistic about the utility of the futile treatment; because parents feel more dependent on the possible good results of the futile treatment. This is a critical issue in medical futility, where professionals argue that the treatment is not beneficial, whereas parents want to try every existing treatment for their child. The desperate attitude of parents is mostly influenced by the relatives. This collective attitude of parents and relatives may lead to the usage of scarce medical resources, while these resources are very costly in "we societies."

However, the scenario can be quite the opposite in Turkey today. The collective unrealistic approach of parents and relatives mostly leads to the refusal of a crucial treatment for the child. According to the Vaccine Workshop organised in March 2018 in Ankara, the number of parents refusing to vaccinate their children was around 100 in 2010, in 2016 there were over 10,000 parents who refused vaccines, while in 2017, the number reached 23,000.[18] On the other hand, as a result of the successful studies of health professionals and parents' sensitiveness about child vaccination, the vaccination rate had increased to 97% in 2015, whereas it was 77% in 1993.[19] Following 2015, the rate of child vaccination refusals by parents in Turkey increased again. The grounds for child vaccination refusals are based on the 1998 research of Wake-

[18] *Turkish News on Vaccine Refusal.* https://www.ntv.com.tr/saglik/turkiyede-asi-karsitligi-artiyor,whwwVUWvYkmE6pdRUWbf3A (accessed October 13, 2018).

[19] S. S. Gulcu, S. Arslan, "An Up-to-date Review: Vaccination Practices in Children," *Journal of Duzce University Health Sciences Institute* 8 (2018): 34-43. (in Turkish).

field, which claimed that vaccines cause autism, attention deficit disorders, hyperactivity disorders, neurological disorders, and some immune system diseases.[20] Although these published research results were retracted because no causal link could be established between vaccines and autism, as the data were insufficient, it nevertheless influenced parents in Turkey and resulted in the organisation of several anti-vaccination campaigns. The Turkish media discussed a case related to a child vaccination refusal in 2015, which was well-known by the public. After the local courts' decision to remove the guardianship from the parents as they refused to have their child vaccinated, the parents appealed to the Supreme Court. The Supreme Court decided that vaccination cannot be regarded as a foreseeable medical intervention and, furthermore, the law does not provide a basis for the inevitability of the vaccination program.[21] With the driving force of such legal decisions, the effect of the internet and social media became more powerful in encouraging patients to raise their voices, while legal regulations (including penal sanctions for physicians) increase the tendency for defensive medicine. Beyond a doubt, it is ominous to go back to the scientific progress and public awareness steps. Unfortunately, in the current situation, parents deny their children the benefit of settled precautions, such as vaccination protocols.

In Turkey, the feelings of parents and relatives are directly affected by the negative perceptions on health-related issues. Therefore, the disagreement between the physician and the parents, who are under the influence of relatives, need to be solved by different approaches. In "we societies," individuals unconsciously accept that the common rules and decisions are more important than their individual decisions. However, this invisible social pressure can also make parents more willing to be part of the decision-making process for their child suffering from a disease and also share the responsibility with the physician; rather than with their relatives. If there is a disagreement between the physician and the parents in "we societies," giving the parents the right to disagree

[20] T. S. Sathyanarayana Rao, Chittaranjan Andrade, "The MMR Vaccine and Autism: Sensation, Refutation, Retraction and Fraud," *Indian Journal of Psychiatry* 53 (2011): 95-96.

[21] *Turkish Press Release by Supreme Court,* http://www.anayasa.gov.tr/icsayfalar/basin/kararlarailiskinbasinduyurulari/bireyselbasvuru/detay/53.html (accessed November 3, 2018).

first may remind them of the beauty of conflict. As a result, the parents, who are not used to have conflicts with their relatives, may take the conflict with the physician as an opportunity. The agreement resulting from the conflict may become powerful enough to change the social pressure in such societies.

Parents who seek a treatment for their child usually find themselves in a desperate position. For the best interests of their child, they need to learn the disease agents, symptoms, the possible treatments, as well as the physicians they have to contact. They feel desperate because they gather information from very different sources which can cause confusion. Parents call on different sources during the information gathering process. These sources include: other parents whose children are suffering from the same disease; the Internet; social media; social environment; relatives; physicians and the mass media.

The Internet is mostly used for gathering information on the disease. It is very convenient and is generally the first source used because whenever parents want to learn anything about the child's disease, the Internet is instantly available via their cell phones. The European citizen's digital health literacy report shows that the majority of the population, around 59%, used the Internet to search for health information in 2014.[22] In Turkey, the situation is not very different, and according to the Turkish Statistical Institute, 66.8% of the population used the Internet in 2017.[23] Furthermore, even information not published by the media may be accessed on the Internet. However, it is not easy to distinguish between fake and real news online. Social media is the other very effective way to gather information. As fake news rapidly spreads on the Internet, the misinformation spread on social media is just as rapid. The mistake parents mostly do is generalizing every kind of information found online about the disease afflicting their child. Generally, such misinformation has more impact on parents than the information received from the physician. On the other hand, social media ignites parents' superstitious beliefs and their unrealistic approach to the disease of their child. As a result, parents may become more pessi-

---

[22] *EU Digital Health Literacy Report* (2014),
http://ec.europa.eu/commfrontoffice/publicopinion/flash/fl_404_en.pdf (accessed October 13, 2018)

[23] Turkish Statistical Institute, *Statistics on the Number of Internet Users in Turkey,* http://www.tuik.gov.tr/HbPrint.do?id=24862 (accessed October 13, 2018).

mistic about the entire situation. The social environment, the relatives are the other effective information sources in "we societies." In fact, the social environment and the relatives are more powerful and effective than all the other information sources. Parents rely on these judgments and even if the information they gathered is promising for a successful treatment, they may refuse a crucial treatment for their child on the basis of these influences.

The information coming from fake news and the pressure of the social environment are not the product of analysing and reasoning. Physicians possess the medical knowledge and learn about the patient's values and beliefs, then become the formulator of the final decision.[24] The formulation of the final decision does not exclude the effect of the relatives and the attitudes of the parents. However, physicians importantly exclude any misinformation from the final decision.

V. Refugee and migrant children in Turkey

Turkey is hosting the highest number of refugees from Syria, a total of 2,854,968 people. Only a small proportion of the total refugee population, i.e. 260,320 people is living in refugee camps, as the majority is trying to stay out with. 1.28 million refugees are under the age of 18, and 490,000 of them are school children – these being approximately 44% of the Syrian refugee population (in other words, about half the population is under 18 years of age). Most are staying in the cities located on the Syrian border: Kilis, Hatay, Sanliurfa, and Gaziantep. However, the largest refugee population is in Istanbul, comprised of almost 550,000 individuals.[25]

The biggest health challenge faced by refugee children is the communication difficulty with health professionals. These children and adolescents have language obstacles causing problems related to setting an appointment at a medical center. Furthermore, even if they managed to set an appointment, the cultural differences make the communica-

---

[24] B. S. Gerber, A. R. Eiser, "The Patient-Physician Relationship in the Internet Age: Future Prospects and the Research Agenda," *Journal of Medical Internet Research* 3 (2001): 2-10.

[25] *Türkiye'de 'Kayıp bir Kuşak' Oluşmasını Önlemek,* (2017), http://www.unicef.org.tr/files/bilgimerkezi/doc/Children%20of%20Syria_01.2007_TR.pdf (accessed October 13, 2018).

tion with medical personnel problematic. The conditions are even more difficult for refugee parents because both they and their children are in need of medical care. Both parents and children belong to a vulnerable population, which makes parents more desperate and pessimistic. The medical care for refugee children is mostly needed for infectious diseases. However, it was noticed that psychiatric disorder diagnoses, especially among male children, are increasing in the outpatient clinics for refugees. There may be two reasons explaining this situation: first, refugee parents are not seeking medical care for the psychiatric problems of their female children, thus disregarding their mental health problems; or second, the male children are more extrovert as a result of their nature and thus their psychiatric problems attract parents' attention more.[26] Whatever the reason, in both cases, the parents' role in the decision-making process seems blurry. It is not entirely possible to talk about the autonomy of the parents. The cultural differences, their vulnerable position in a foreign country, and the language difficulties causing communication problems suppress the autonomy of refugee parents. Physicians may have limited language skills to explain the futility or the cruciality of a particular treatment to refugee parents who cannot speak Turkish or even English. As a result, the best interest of the refugee child is mostly determined by physicians.

VI. Parents refusing children's treatment

Parents have the right over their children as guardians. The scope of this right is very wide, so much so that it is usually understood to include a right to refuse crucial treatment for their children. However, according to the Constitution, as legal guardians, parents have the responsibility to ensure the fundamental rights of the child including the right to health. The child has to be under guardianship until the age of 18, in as healthy a condition as possible. If it is determined that the guardianship responsibility is not fulfilled, the court has the power to remove guardianship from the parents. If the guardianship is revoked from both mother and father, a different guardian may be appointed.

[26] S. Oguz et al, "War and Children: Effect of Syria Civil War on a Pediatric Emergency Department, 750 km Away from the Border," *Journal of Pediatric Emergency and Intensive Care Medicine* 3 (2016): 135-139.

As a concluding remark for vaccination and other crucial treatments for minors, there must be some limitations for parents' refusal. The following may be acceptable circumstances for refusal:

- When alternatives of the treatment exist
- Treatment examined during a clinical research
- Sufficiently unproven experimental treatments
- Treatment with a high mortality rate in the presence of a fatal disease.

The criteria above give parents the right to justifiably refuse as it reduces "the possible harms the treatment would cause to the child." However, physicians have to explain to them the other possible harms when refusing treatment, such as that: [27]

- The refusal may put the child at serious harm, and the harm may be about to happen
- The refused medical intervention has been proven to be efficient
- The possible benefit of refused treatment is higher than the parents' preferred option.

These criteria may help physicians to discourage parents from refusing treatment. Furthermore, although the parents are rational individuals capable of prioritising the options for the best interests of their child, there may be other factors affecting their final decision, such as relatives. Therefore, diagnosing the reasons for the refusal may be helpful to reduce the disagreement between the physicians and the parents. If the reason for refusal stems from the relatives, the physician needs to change the perception of the relatives as well as the viewpoints of the parents with the information they need. In Turkey, the concept of patients' rights is perceived in such a way that it may increase the attacks against the physician as a result of the relatives of the parents displaying threatening attitudes. Surely, in order to achieve the best interests of the child, all the parties involved must resolve their disagreement. More data on paediatric treatment cases in Turkey is needed, in terms of the attitudes of the physicians and the relatives of the child including his/her parents.

---

[27] R. McDougall, L. Notini, "Overriding Parents' Medical Decisions For Their Children: A Systematic Review Of Normative Literature," *Journal of Medical Ethics* 40 (2014): 448–452.

VII. Conclusion

We generally assume that the physician knows the best. The important point we overlook in this assumption is the necessity for physicians to know the parents' values too. Furthermore, physicians need to analyse "what would the parents choose?" before explaining to them the medical conditions of their child.[28] This implies guiding parents, nudging them and creating a "choice architecture" designed to influence parents' behaviors in a predictable way without restricting. In Turkey, a country with many different cultural backgrounds, such an analysis is extremely important.

After the physician's analysis, a shared decision-making between physician and parents should take place. Physicians and parents may have different values and knowledge. This means that both sides would have to try to understand the reasoning of the other side, and thus come to a shared decision. An unsuccessful sharing during the decision-making process may cause wrong assumptions and lead to conflicts between the physicians and the parents. Physicians have to listen to parents carefully, try to understand what they are saying, what they mean and what they need.[29] Sometimes, physicians need to understand what parents cannot say when they actually want to tell something different. In some cases, physicians should let parents disagree and show them that disagreement is an opportunity to reach a steady agreement level. These sensitive approaches may enable physicians to overcome wrong assumptions.

Finally, the socio-economic status, geographic location, the type and extent of education, as well as family traditions are all components of culture. Cultures are formed of different levels of these components; which make them different from each other. However, culture is not a stable norm, thus the interaction between different cultures changes the levels of the components such as traditions or extent of education.[30] Therefore, the interaction between the physicians and the parents from

[28] D. Brudney, "Practical Wisdom, Rules and the Parent-Pediatrician Conversation," *Pediatrics* 142 (2018): 193-198.

[29] J. D. Lantos, "Tell Parents the Truth, but Tell It Slant," *Pediatrics* 142 (2018): 199-204.

[30] S. F. Derington et al, "Cross-Cultural Interactions and Shared Decision Making," *Pediatrics* 142 (2018): 187-192.

different cultures affect the components of cultures of both sides. Disagreement is a process of cultural interaction, and this is rather an experience for physicians and a learning process for parents. The key element here is the limited time for the child's survival; thus the experience process for physicians and the learning process for parents are both very precious if they do not exceed the time needed for the child's survival.

Based on the discussion above, if situations similar to Charlie Gard's and Alfie Evans' were to arise in Turkey, the decision-making concerning the withdrawal of their life-sustaining treatment would involve additional parties, namely relatives. However, physicians would have more authority to decide on the continuation or otherwise of life-sustaining treatment even before a court is approached. A similar scenario happened in 2015, in Turkey. A preschool girl R went into a liver coma and needed a liver transplant. R's father was determined to be a suitable donor. However, just after the operation start, both the relatives of the father and R's mother objected to the operation. Physicians paused the operation for about an hour, to re-emphasise to them the importance of the operation and that the child's survival chance is approximately 70%; which cannot be ignored. However, both sets of relatives thought that the child had no chance of survival and the father was taking a very high risk with the operation. After the insistent attitudes of the patient's relatives, the physicians terminated the operation at that phase. At that phase, the only alternative physicians have was to wait for an organ to be transplanted from a cadaver. Unfortunately, R died two days after the unfinished operation. R's case is an example of medical futility in paediatrics in Turkey.[31] Although Turkish law gives more authority to physicians to decide on the continuation of the treatment of a child; the insistent, and most of the time threatening, attitude of the child's parents and their relatives intimidate the physicians. Although the law allows physicians to decide the best interests of the child according to their medical knowledge; in the practical and real environment, physicians also need to have appropriate conditions to make free decisions about the chid; without remaining under pressure.

[31] C. H. Guvercin, B. Arda, "Parents Refusing Treatment of the Child: A Discussion About Child's Health Right and Parental Paternalism," *Clinical Ethics* 8 (2013): 52–60.

Chapter 16

If We Can, Must We?
Just Whose Best Interests are We Talking About?
Perspectives from the USA

Vincent F. Maher
Iona College, New York, USA

I. Introduction

It seems that not infrequently in life, things come in threes. Sneezes and death come to mind immediately. Such have been the situations addressed in this chapter. The three tragedies that occurred proximate in time to one another in different parts of the world, in medically and socially sophisticated societies where there are well-established medical and ethical interventions and parameters, demonstrated in painful ways the raw underbelly of medical care. Patient/medical futility, parental wishes, rights and responsibilities, differing and evolving views of beneficence, and the standing to make decisions that are specific to an individual or an individual's family member fuelled by mutual mistrust of the constituent groups of one another, and the overwhelming role of social media all joined to form perfect storms. The cases of Charlie Gard and Alfie Evans have been outlined elsewhere in this book. In the USA, around the same time, a comparably painful case had unfolded. That case is the primary focus of this chapter. What follows is a brief summary of the Jahi McMath case[1] followed by a comparison of the three cases, and thereafter what appears to be evolving needs on the parts of the involved parties. Differences in national laws, perceptions

---

[1] For a more detailed discussion of the case, see V. Maher, G. Priovolos, E. J. Maher, K. Molloy, "The Girl Who Lived: The Case of Jahi McMath," *Journal of Leadership and Management* 4:14 (2018): 238-245. See also V. Maher, G. Priovolos, E. J. Maher, "When Technology Complicates Management: The Case of Jahi McMath" (2014) Cambridge Conference Business & Economics.

of parental preferences, and medical control, and of interpretations of ethics and legal standards contribute to the confused discussion.

II. Jahi McMath[2]

In November 2013, thirteen year old Jahi McMath ("Jahi") was suffering from sleep apnoea secondary to morbid obesity. She agreed, through her parents, to undergo corrective surgery. Nonsurgical methods commonly used to correct morbid obesity and sleep apnoea had been attempted previously and failed.

Surgery for sleep apnoea can range from the simple to the complex. In the simplest iteration, surgery involves a tonsillectomy – removal of the tonsils. Whilst this may be a fairly straightforward procedure in children, it becomes more complicated in the case of teens and adults since problematic intraoperative and post-operative bleeding can occur with greater frequency. The procedure to correct sleep apnoea can become more complex as one checks items on the *a la carte* surgical menu of interventions that include uveloplasty, paletoplasty, adenoidectomy, uvelectomy, phararyngopaletoplasty, and so on. In lay terms, we are talking about reconstruction of some or all of the soft tissues of the posterior mouth and airway.

In this particular case, the more extensive multisite and invasive surgery was anticipated by her surgeons, which led to the procedures they performed. Based on subsequent media reports and court filings, Jahi came through the surgery in a satisfactory fashion and was transferred for routine monitoring to the post anaesthesia care unit (PACU). It was reported that she was responsive and interacting with her mother and grandmother and was even asking for an ice popsicle. Following her stay in the PACU, and as previously planned, she was transferred to a routine post-operative paediatric bed for overnight monitoring. It was anticipated that she would be discharged the following day.

Something went very wrong.

Jahi's grandmother, described in various media outlets as a "veteran" nurse, is reported to have noticed that Jahi was bleeding from the mouth and nose. Whether Jahi had been asleep or awake, or awake and talking, or awake, talking and eating at the time has not been reported.

---

[2] *Winkfield v Children Hospital Oakland et al (2013), California Northern District Court Oakland Division*, Case No: C13-5993.

Unconfirmed reports (because case records remain closed due to pending litigation) suggested that contrary to medical directives, Jahi had been given food and drink allegedly by her grandmother, the only visitor at the time, which proximately caused the post-operative bleeding. Whatever and whoever it was that started the problem does not really matter: Jahi began to haemorrhage. Rescue efforts ensued, but the public was not informed as to whom these efforts were performed by or indeed for how long. Some sources have stated that, regardless of the cause of the bleeding or of what may or may not have constituted a proper staff response, the grandmother attempted to suction the blood from Jahi's mouth.[3] Even as blood was cleared, bleeding was exacerbated. Jahi was without oxygen for some time and, although "successfully resuscitated," she was transferred to a critical care unit in extremely critical condition. A few days later, multiple neurologists concluded, based on the period of anoxia and of standard contemporary neurological testing in this context, she was brain dead. The case should have concluded at this point. It did not.

III. The problem with definitions

There is what science says, and then there is what a person believes to be true. So it has always been: so it continues to be. If these two things are not congruent, there will be, at a minimum, conflict. While there should be individual considerations for cultural and religious practices when a person dies, serious social and policy issues arise when there is a conflict in the definitions of life, death, proper/improper medical treatment, and medical futility or the understandings thereof. These issues have been dealt with since the time of Hippocrates and more recently in the mid to late 20th century. For example, the rise of modern medicine prompted Pope Pius XII to speak out about extraordinary medical interventions in 1957 saying that there was no moral obligation to undertake treatment where there was no hope of recovery.[4] Ethicist Gerald Kelly, SJ said on behalf of the Catholic Health Association of

[3] C. Johnson, "On Brain Death and the Jahi McMath Story," https://www.chrisjohnsonmd.com/2014/01/21/on-brain-death-and-the-jahi-mcmath-story/ (accessed March 30, 2019).

[4] Pope Pius XII, "The Prolongation of Life," in *Acta Apostolica Sedi* (Vatican City: Vatican City Press, 1957).

the USA and Canada, that "extraordinary medical means are all medicines, treatments and operations, which cannot be used without excessive expense, pain or other inconvenience, or which, if used, would not offer a reasonable benefit of hope."[5]

Since that time, the professional medical and ethics literature mostly concludes that while recognising medicine's role in decision-making and the application of the ethics progeny of the *Quinlan*,[6] *Brophy*[7] et al cases of the 1970s and 1980s, where patients and/or their families refused continued medical treatment, medical diagnosis and treatment interventions should generally take precedence where there is conflict and particularly where intervention is deemed futile.[8]

These three cases of Jahi McMath, Charlie Gard, and Alfie Evans not only challenged well-established medical, ethical, and even judicial presumptions, they threw the presumptions into the light of public scrutiny and opinion via social media in spite of warnings to the contrary from authors such as the NYS Appellate Division in 1989 in the *In Re Klein* case wherein the court pointedly chastised non-party interventionists[9] and, more recently, Wilkinson who warned of the inappropriateness of social media intervention.[10]

Jahi had been pronounced brain dead by three different teams of medical professionals not long after the incident described above.[11]

---

[5] G. Kelly, SJ, "Commentary on the Prolongation of Life by Pope Pius XII," in *US and Canadian Bishops Health Care Conference* (Washington, DC: Catholic Health Association of the US and Canada, 1958).

[6] *Re Quinlan* 55 A.2d 647 (N.J. 1976).

[7] *Brophy v New England Sinai Hospital* 398 Mass. 417, N.E. 2d 626 (1986).

[8] L. J. Schneiderman, "Still Saving the Life of Ethics," in *The Hasting Center Report* (The Hastings Center: Wiley Online Library, 1990). R. L. Schwartz, "Autonomy, Futility and the Limits of Medicine," *Cambridge Journal of Ethics (Cambridge Quarterly of Healthcare Ethics)* 1 (1992): 159-164. B. A. Brody, A. Halvey, "Is Futility a Futile Concept?" *Journal of Medicine and Philosophy* 20:2 (1995): 123-144.

[9] *In Re Klein,* NY Suppl 538 (1989): 274-6 (Appellate Division, NY State, NY).

[10] "How Can We Achieve Greater Balance in Future Cases Like Charlie Gard, Asks Expert?" *British Medical Journal* (2017), https://www.bmj.com/company/newsroom/how-can-we-achieve-greater-balance-in-future-cases-like-charlie-gards-asks-expert/.

[11] R. Arbour, "Brain Death: Assessment Controversy and Confounding Factors," *Critical Care Nursing* 33:6 (2013): 27-46.

Despite this, her family did not believe she was dead because she still retained a heartbeat.[12] The heartbeat was likely caused by the mechanical assistance that she was receiving. If Jahi were removed from mechanical and pharmacological assistance, her heart would eventually stop beating because the integrative functions performed by the brain in this regard had been destroyed and were dead.

Jahi's family was not expecting her to die from the elective surgery she underwent. In this instance, they could be experiencing guilt because they facilitated and/or allowed this condition (morbid obesity) to develop along with the eventual need for surgical intervention. There was no time for the family to prepare for this loss and that, too, made it difficult for her family to come to terms with her death. Her mother, Nailah Winkfield, stated that she would only accept that her daughter was dead once her heart stopped.[13] This is, sadly, a not uncommon response even though it has been addressed in the United States Uniform Death Determination Act of 1968, meaningfully operationalised even further, for example, by the Texas Health and Safety Code, Public Health Provisions, the Advance Directives Act of 1989 and directly applied in the 2005 case of Sun Hudson who was born with a fatal congenital defect and whose breathing tube was removed against the specific wishes of his mother because continued care was deemed by his medical team to be futile.[14]

Guilt, denial, and fear of reprisals for potentially negligent acts can also be profoundly motivating forces on the part of the family. Let us, too, not underestimate the fact that conflict between parents and caregivers may be related to the fact that parents may subconsciously perceive their children as an extension of themselves that live on after parents are gone. Letting go of a child in a terminal situation such as

---

[12] Ibidem.

[13] E. B. Andrade, D. Ariely, "The Enduring Impact of Transient Emotions on Decision Making," *Organizational Behavior and Human Decision Processes* 109:1 (2009): 1-8.

[14] R. D. Feigin, L. B. McCullough, "The Tragedy of Sun Hudson: There Was No 'Life or Death' Decision to Make" (27 March 2005), https://www.chron.com/opinion/outlook/article/The-tragedy-of-Sun-Hudson-1947366.php (accessed April 2, 2019).

these would mean relinquishing the one tightly held little piece of immortality.[15]

The medical staff of the treating hospital, Children's Hospital Oakland, were able to quantitatively demonstrate that Jahi was clinically brain dead. A non-affiliated independent third party, court appointed medical team also agreed, following extensive testing, that she was brain dead and that mechanical assistance could be removed. Since three different medical teams independently confirmed brain death, the hospital acted in its right to not perform medical procedures on a person who was no longer alive. It should be obvious that insurance companies will also not pay for the continued care and treatment of a now dead body. On average, in the USA, keeping a person in the intensive care unit (ICU) on mechanical assistance and life support can cost at least $5,000 a day.[16]

Ideally, most experts agree that decisions, especially major decisions, should never be made in an emotional environment because "incidental emotions can influence judgment and decision-making."[17] Emotions can influence decisions long after the emotion has dissipated. Even knowing that, people tend to act and make decisions as they have in the past, or to support decisions that have been made based on that original emotional decision. Hospitals have ethics committees so that cases like these – where there is or can be faulty decision-making – can be addressed while adhering to established ethical principles that govern medical ethics and include dispositive values such as autonomy, beneficence, non-maleficence, veracity, equity, confidentiality, and justice.

In Jahi's case, the hospital accepted the clinical diagnosis and the court ruling, and pressed its point. Specifically, the hospital publicly declared her dead in early December, permitted the cessation of fluid and nutritional support, and agreed with the decision to remove life support.

The family, through counsel, blamed the hospital for what happened. They persistently fought the decision to remove life-sustaining treatment, and insisted on her immediate transfer to another "sympathetic facility." Eventually a mediator assisted the parties to come to

---

15 M. Gervasi, interview by V. Maher. *Thoughts on Parents of Seriously Ill Children* (January, 2019).

16 R. Arbour, "Brain Death."

17 E. B. Andrade, D. Ariely, "The Enduring Impact of Transient Emotions."

resolution, to wit: having been declared dead and the hospital having issued a death certificate in early December, her body, still on life support, was handed over to the custody of the county coroner in early January. The coroner, in turn, released her body to her family along with medical records and a death certificate. Jahi was transferred from her California hospital to an unnamed facility (ultimately determined to be St. Peter's Hospital, New Jersey) that had agreed to accept her and to keep her on life support. New Jersey law permits a family to refuse to accept a diagnosis of brain death if such a diagnosis conflicts with the religious beliefs of the family.[18]

Since the time of transfer, a media blackout was in place other than to say that Jahi had received a tracheostomy for long-term ventilator and pulmonary management, a gastrostomy tube for nutrition purposes, and that she was ultimately transferred to an apartment in Somerset, New Jersey that was supplied with the equipment necessary for her family to sustain her. In 2018, without any doubt or argument, Jahi succumbed following emergency abdominal surgery for abnormal bleeding.[19] Were 5 years of such care and treatment truly in the best interests of this child?

IV. Decision-making

When patients' clinical situations deteriorate, it is important to realise and acknowledge that in critical care decision-making, the use of clinical algorithms is standard and accepted medical practice. The aim is to remove subjectivity from the equation, to ensure thoroughness, to eliminate useless or inappropriate testing, and to ensure medical equity for other patients in comparable situations. Even without seeing a medical record, people accustomed to clinical decision-making would expect to see certain behaviours on the part of doctors and nurses. What is also important to these scenarios is that even in the face of everything being carried out precisely and on time, a patient can deteriorate and die. Such was believed to be the case here. For example, Jahi's post-resuscitation clinical signs, symptoms, and responses would have compelled a consultation by a neurologist. A neurologist is not a member of

[18] Section 6, New Jersey Declaration of Death Act 1991.

[19] D. Goldschmidt, "California Teen at Center of Brain-Death Controversy, Has Died," CNN, 29 June 2018.

the surgical team. A neurologist enters the frame, evaluates available data, conducts a physical examination, renders a diagnosis and leaves. That diagnosis results in a clinical intervention consistent with the diagnosis. In this case, the diagnosis was death: the clinical decision and intervention were to discontinue life support. Although this is all rather obvious and routine to healthcare professionals, the same was not the case for the family.

Parents are programmed to hope. Recent advancements in research and technology have led to better health outcomes and the improved ability to prolong life.[20] Consequently, a failure of these advancements to cure a loved one can come as a surprise and disappointment. Subsequent decision-making relating to the withdrawal or the withholding of treatment can be wrought with difficulty and often full of emotion.

A family whose teenaged daughter was alive a few days earlier and who underwent elective surgery done on a routine basis for this sort of problem and who was now in a coma but whose heart was still beating is going to have a difficult time listening to someone telling them that their child is dead and that her life support machines need to be shut off and removed. In such dramatic and emotionally volatile circumstances, healthcare professionals are going to be hard-pressed to insist that this discontinuance happen immediately. Unfortunately, what drives the well-intentioned act of forbearance, or fear, then becomes a nightmare for all as the heart refuses to cooperate and to cease beating.

In Jahi's case, available technology kept a heart beating. If the technology did not exist or were not available for use depending on things like geography, financial and technological soundness of the facility, we would not be discussing this or any of these cases for that matter in the first place. Technology frequently exists either *in situ* or in a place to which transport can frequently be arranged. Hope rises in direct proportion to availability and use.

V. The role of futility

Hippocrates anticipated the medical dilemma of futility when he advised physicians to "refuse to treat those who are overmastered by their

---

20 I. O'Brien, A. Duffy, E. O'Shea, "Medical Futility in Children's Nursing: Making End of Life Decisions," *British Journal of Nursing* 19:6 (2010): 352-356.

diseases, realising that in such cases medicine is powerless."[21] Hippocrates' admonition has been followed since whether medical practitioners were treating plague victims, soldiers in times of war, or individuals victimised by the ravages of natural disasters. In the USA, contemporary medical personnel have grappled with problems of futility and near futility in the care and treatment of patients for decades.[22]

It is not unreasonable to ask the following question: given the length of time and the amount of literature and law that exists on this issue, how is it even possible that we are still discussing these issues? The answer seems to be threefold at least in terms of relevance to the issue of the best interests of the child and of parental preferences, rights and responsibilities: 1) advanced medical knowledge and contemporary technology available to providers and patients especially in developed nations that creates an expectation that knowledge and technology will be used without question or restraint in any and all circumstances; 2) consumer expectations/refusal of anything other than perfect results even in complicated or futile medical care environments and the attendant ability of the aggrieved to resort themselves to social media or the courts to plead their denied cases; and 3) an apparent conflict of laws among federal laws (Uniform Declaration of Death Act of 1981, the Federal Rehabilitation Act of 1973 and 1978, the American with Disabilities Act 1990), federal law (*Cruzan v Director*, DMH 497 US 261 (1990)), and state laws (the New Jersey Declaration of Death Act of 1991).

VI. Charlie Gard

Charlie Gard suffered from a genetic disorder that had no treatment and was fatal. In the Gard case, however, there was the remote potential for intervention based on experimental studies conducted at a number of world class medical centres including the team led by Pro-

[21] S. Reiser, "Hippocrates," in *Ethics in Medicine: Historical Perspectives and Contemporary Concerns*, ed. Stanley Joel Reiser, Arthur J. Dyck, William J. Curran (Cambridge: MIT Press, 1977), 6-7.

[22] S. H. Miles, "Informed Demand for 'Nonbeneficial Medical Treatment'," *New England Journal of Medicine* 325 (1991): 512-515. R. D. Truog, A. S. Brett, J. Frader, "The Problem with Futility," *New England Journal of Medicine* 326 (1992): 1560-1564.

fessor Hirano of the Neurological Institute at Columbia University Medical Center, New York. Discussions between Professor Hirano and the Great Ormond Street Hospital medical staff who cared for Charlie made it clear that Charlie could receive the experimental treatment only if there were no brain damage. Logistics of various kinds intervened and, by the time scans were done, information was shared, and the child was examined by Professor Hirano, there had been a delay such that Professor Hirano determined that the experimental treatment would not work due to what had become extensive brain damage.[23] The parents refused to accept this determination and created a crowd-funding page[24] to transport Charlie to the USA for treatment by Professor Hirano that had already been determined by him to be of no use. They also launched legal appeals to the English courts and the European Court of Human Rights. The legal appeals failed.[25] Meanwhile, President Trump said he would grant legal resident status to Charlie and his family in order to come to the USA to receive treatment.[26] This rescue mission did not happen as Charlie rapidly deteriorated, was placed in hospice, and died within a day of placement.

VII. Alfie Evans

In 2018, Alfie Evans was emergently admitted to hospital and was diagnosed with an unnamed but disastrous and fatal progressive neurological condition. The hospital wanted to turn off the life support based upon his medical deterioration. His parents objected and sued when the medical team refused their demands to continue medical intervention. This case also went to the English High Court where Mr Justice Hayden upheld the 1989 Children Act that makes clear that when a child's best interests are in contention, the state can and should intervene. Mr

---

[23] R. Jayaram, "Would Charlie Gard Really Have Survived if He'd Been Treated Sooner? Here's the Sad Truth," https://www.independent.co.uk/voices/charlie-gard-gosh-hirano-would-he-have-survived-a7861531.html (accessed May, 15, 2019).

[24] Charlie Gard #charliesfight, https://www.gofundme.com/please-help-to-save-charlies-life (accessed April 30, 2019).

[25] See the Inroductory chapter of this book.

[26] A. Dewan, S. Elwazer, "US Congress Takes Step Toward Giving Charlie Gard Residency," https://edition.cnn.com/2017/07/19/health/charlie-gard-us-residency/index.html (accessed April 15, 2019).

Justice Hayden said that continued treatment of Alfie was inappropriate and that doctors could stop life support against his parents' wishes saying the child required "peace, quiet and privacy."[27] The parents refused this ruling and proceeded further with appeals within and beyond the UK that were also denied. The European Court of Human Rights specifically said it could not hear the case because it found no violation of Alfie's human rights. Mr Evans, Alfie's father, who had become quite adept at social media, then turned to the Pope for help. Italy granted Pope Francis' request and conferred citizenship on Alfie in hopes he could be transferred to the Bambino Gesù Hospital for underdetermined medical and surgical intervention to prolong his life. Pope Francis met Mr Evans privately, blessed him and thereafter tweeted his support for the family's decision to continue treatment: "I renew my appeal that *the suffering of his parents may be heard* and that their desire to seek new forms of treatment may be granted."[28] It would appear that even among popes there is disagreement. Pope John Paul II recapitulated Pope Pius XII, when he wrote:

> Certainly there is a moral obligation to care for oneself and to allow oneself to be cared for BUT this duty must take into account concrete circumstances. It needs to be determined whether the means of treatment available are objectively proportionate to the prospects of improvement.[29]

The emotional suffering of the parents is not part of the equation. Or is it?

And there, in a nutshell, is the problem. The decisions on Alfie, Charlie, and Jahi, as individual people, had devolved to a social media circus about who gets to decide what, whose values should be given priority where there is a conflict of values, and to the emotional suffering of the parents. Many people on social media seemed to have an opinion that they felt entitled to not only share but to which medical and legal teams were bound by to intervene!

---

[27] *Alder Hey Children's NHS Foundation Trust v Evans, James and Evans* [2018] EWHC 308 (Fam) para 62.

[28] "Pope Francis Renews Appeal for Alfie Evans," *Vatican News*, 24 April 2018, https://www.vaticannews.va/en/pope/news/2018-04/pope-alfie-evans-twitter-appeal.html (accessed March 30, 2019).

[29] Pope John Paul, *Evangelicum Vitae* (Vatican City: Vatican Press, 1995).

In social media and in the referenced political interventions, standard practice in medical decision-making processes appeared to have been cast aside in preference to the parents, their needs, desires, rights and responsibilities pertaining to the best interests of their children. Is there in fact a medical, ethical, and legal practice shift and, if so, what is the origin of such a shift? No, there is not. At least, not officially. In the UK, for example, the answer is found in the 1989 Children Act and in the 1985 *Gillick* case wherein Lord Scarman stated that

> There is here a principle which limits and governs the exercise of parental rights of custody, care and control. It is a principle perfectly consistent with the law's recognition of the parent as the natural guardian of the child; but it is also a warning that parental right must be exercised in accordance with the welfare principle and can be challenged, *even overridden* (emphasis added) if it is not.[30]

In the USA, the Hudson case in Texas would advocate on behalf of this principle and of attendant reasonable medical practice and legal non-intervention.[31]

VIII. The impact of social media

It seems appropriate at this juncture to address the role of social media especially wherein those with no standing seem not only obliged to speak up on behalf of the perceived victims but also to have their opinions followed. There has always been a media circus surrounding controversial medical interventions. There are also judges, for example, who have blatantly told those who would insert themselves into processes that do not directly concern them to desist. For example, in a 1989 New York case, *In Re Klein*, wherein Martin Klein sought guardianship of his wife in order to permit physicians to perform an abortion on her when she had been left comatose after a motor vehicle accident. It was the opinion of Mrs Klein's treating physicians that an abortion would help to awaken Mrs Klein from her coma. This decision was challenged by third party individuals. Pro-life activists John Short and John Broderick who heard of this proposed procedure intervened with

---

30 *Gillick v West Norfolk and Wisbech Health Authority* [1985] 3 WLR 830.

31 R.D Feigin, L.B. McCullough, "The Tragedy of Sun Hudson."

legal challenges to have Broderick named as guardian of Klein's foetus and Short named as guardian of Mrs Klein in order to prevent the abortion. Meanwhile, Operation Rescue run by Broderick, somewhat akin to Alfie's Army, staged anti-abortion protests at the hospital where Mrs Klein was being treated. Protesters were arrested for trespass.[32] Following the initial court's determination awarding guardianship and decision-making authority to the husband, the case went for an immediate review by an appeals court. There, the five-judge appellate panel unanimously affirmed the lower court's decision and took the opportunity to pointedly call the abortion foes "strangers who had no place in the midst of this family tragedy."[33] From there, continued appeals by the anti-abortion advocates went to the New York Court of Appeals which refused their request, and then to the US Supreme Court which also refused to intervene.[34] The abortion was performed: Mrs Klein awakened and has since recovered.

IX. Discussion

Alfie's, Charlie's, and Jahi's fates had already been determined by pathophysiology. No medical intervention of any kind was going to reverse those courses. The fundamental issue of the best interests of the child standard becomes one of implementation: we all seem to know what it is and what to do but are less clear about when and how it should happen. At what point do the medical decision-makers stand back and give parents room and at what point do they push the conclusions that they are trained and licensed to diagnose and treat, or not treat? Parenthood confers many things. It does not confer medical knowledge or legal credentials any more than it confers athletic, musical, electrical, or mechanical ability no matter how much a parent may desire those attributes in their child. Parenthood does not confer absolute rights. At the risk of oversimplification: just as an electrician or a plumber must abide by regulatory codes to promote public safety and health regardless of what a client may desire, so too, must physicians abide by best medical

32 C. Page, *Pro Life Intruders in a Close Call Family Tragedy,* News Report (Chicago: Chicago Tribune, 1989).

33 *In Re Klein.* 1989. NY Suppl 538: 274-6 (Appellate Division, NY State, NY).

34 E. Schmitt, *Appeal on Abortion Refused: Woman in Coma is Readied,* News Article (New York: The New York Times, 1989).

practices as commonly accepted in the profession to maintain public health, equity, and safety even where those practices may not be embraced by patients, their families, or even well-intentioned presidents or popes. In the cases of all three minors, the medicine, although complex, was straightforward as were the clinical algorithmic interventions put in place. It should be easily apparent to anyone that the continued care and treatment of these children was prolonging their lives but for whom and for what purpose? Were they going to recover? No. If they were not going to recover, then why were they being treated? For hopes of a miracle? To ease the parents' anguish? Was withdrawal of treatment hastening death? Perhaps, or was it merely letting nature take its course? Was that a bad thing? An evil thing? Was it something that would not be implemented in the care and treatment of other similarly-situated children? Should the technology have been placed in the first place when diagnoses were pending? Should the technology, once placed, have been withdrawn as soon as a diagnosis had been confirmed rather than subject the children to protracted pain and suffering foisted on them by well-meaning adults including their parents? Are the treatments and interventions that these children received going to be put into place with each and every other child who presents for comparable treatment in the future? Surely, ethical practice would compel such interventions by physicians yet one wonders whether ethical practice would demand withholding/withdrawal of treatment at earlier stages? Autonomy does not confer the ability to mandate behaviour by other individuals or communities of interest. In other words, there are limits. Not a word we like to see or live with, but it is there nonetheless. Limits. I cannot be anything I want to be if I lack the skill sets for that desired occupation. I cannot live anywhere I wish without the income to support that, etc. I cannot live if my physiology determines otherwise. So, too, is the case with making decisions in the best interests of our children. We want what is right for them even as we desire a wonderful outcome. The actual, objective facts and anticipated outcome, however, may compel beneficent acts by physicians and others that are contrary to that which we want and desire.

X. Conclusions and recommendations: what is to be done differently?

First and foremost, we need to face the reality that cases like Alfie, Charlie, and even Jahi will continue to occur. In the absence of universal prenatal genetic screening, children with these kinds of congenital anomalies will continue to be born. In spite of best medical efforts, disasters that afflicted Jahi will occur. These realities anticipated, how does society establish parameters that best serve the needs of providers, patients, families, and society when these things happen?[35]

Early diagnosis by a medical team will put into place best contemporary medical diagnostic knowledge and practice, and attendant clinical algorithms of care that already include the medical best interests of the child. If the anomaly is known to be fatal, such as the cases presented herein, then interdisciplinary palliative and hospice care should be provided to the child, and mental health services to the family, from the outset of the child's life until the child's death. Putting "just in case technology" into place is a disservice to the child, the family, and to the integrity of the health care system. In cases such as McMath, early diagnosis and determinations of prognosis, especially in a landscape of changing interpretations of brain death and integrated neurological functionality, by an interdisciplinary team will also put into place, tried and true algorithms of medical intervention including palliative and hospice care designed to maintain the best interests of the child and to realistically differentiate hope with medical futility. Society should not be compelled to sustain or to fund medical treatments based on someone else's religious preferences even if a local law permits a family to reject a medical diagnosis for religious reasons. One does not compel the other. In the USA specifically, the 1st Amendment separates church and state. The 4th Amendment prohibits unreasonable seizure of individuals by the government. These protections are not set aside in the case of individuals by the 14th Amendment.

Secondly, families need to understand that they are free to utilise social media of choice but that the medical care provided to their child will not be determined by the uninformed and uneducated voice of

[35] C. J. Misak, D. B. White, R. D. Truog, "Medically Inappropriate or Futile Treatment: Deliberation and Justification," *Journal of Medicine and Philosophy* 41:1 (2016): 90-114.

public opinion or of well-intentioned but poorly informed preferences or interventions by anyone other than the medical team directly caring for the child. The medical team can and should seek internal and external medical consultations where appropriate and should continue to make decisions that have a direct correlation to likelihood of success, not just temporary survival. This intervention does not diminish the role of the parents in decision-making. It does deal with the patient and not with the parents' pain at the plight of their child. It does, however, recognise that the medical issues at hand are highly complicated and that a session or two in a conference room does not substitute for years of education, training, and the licensed ability by the state of a physician to make appropriate decisions on the one hand or a lifetime of religious belief on the other hand.

Thirdly, the courts should demurrer to hear these kinds of cases in which the parties are "shopping" for an answer they desire and leave the medicine to the physicians just as physicians leave the practice of law to the attorneys and to the courts. This forbearance does not preclude legal motions that are *de novo* and that legitimately need to be heard.

Finally, although all would wish otherwise, miracles appear to be low probability events in these cases. All involved in these tragic cases need to accept the fact that these children are terminally ill and will die. All involved need to accept the best available science underlying the diagnoses, the available and credible technology that can be attempted, and that the care and treatment of these children are provided with well-established clinical algorithms derived from many similar cases in the venues where these cases have occurred and beyond. All involved acknowledge the emotional pain the parents are experiencing and are to offer emotional support to the parents.

The best interests of the child are always paramount. It behooves all of us to make that as easy, as safe, as timely, and as medically and ethically reproducible as possible to all similarly-situated even in the face of parental emotional distress.

TRIVENT

Ethics | in Science &Technology

This book is part of the *TRIVENT Ethics in Science & Technology (TEST) Imprint*. TEST intends to provide insight into the ethical issues and underlying problems in research and innovation. It includes book series covering the wide range of areas from applied philosophy to ethics and bioethics. TEST places a special emphasis on the ethics of new technologies such as digital technologies (AI, Big data) and bioengineering technologies (Gene editing, CRISPR). Besides the ethical issues related to new technologies, TEST also confronts ethical and social issues related to the environment, climate change, medicine, and issues related to research integrity. It also investigates the philosophical and cultural approaches related to the discoveries of new technologies, posthumanism, and transhumanism. Therefore, TEST brings an adequate analysis of the ethical, legal, and social issues of innovative scientific and technological discoveries.

FSC
www.fsc.org
MIX
Papier | Fördert gute Waldnutzung
FSC® C083411

Kolibri 360 GmbH
Ferdinand-Jühlke-Straße 7
99095 Erfurt, Deutschland
produktsicherheit@kolibri360.de